GROUND BASED AIR DEFENCE IN INDIA

Challenges and Opportunities

GROUND BASED AIR DEFENCE IN INDIA

Challenges and Opportunities

Lt Gen (Dr) V K Saxena (Retd), PVSM, AVSM, VSM

United Service Institution of India, New Delhi

PENTAGON PRESS

Ground Based Air Defence in India: Challenges and Opportunities
Lt Gen (Dr) V K Saxena, (Retd), PVSM, AVSM, VSM

First Published in 2018

ISBN 978-93-86618-58-0

Published by
PENTAGON PRESS
206, Peacock Lane, Shahpur Jat
New Delhi-110049
Phones: 011-64706243, 26491568
Telefax: 011-26490600
email: rajan@pentagonpress.in
website: www.pentagonpress.in

Printed at Aegean Offset Printers, Greater Noida, U.P.

This book is dedicated to my granddaughter, Nia

CONTENTS

FOREWORD

Lt Gen Naresh Chand, (Retd), PVSM

Once again Lt. Gen. Dr. V. K. Saxena (VK) surprised me by asking me to write the foreword of his book titled, 'Ground Based Air Defence in India: Challenges and Opportunities'. The surprise was that he already has five books to his credit and this is his sixth book. It will be no exaggeration to state 'that his interest and staying power for academic pursuits is unmatched.' When he asked me to contribute a chapter for this book, my immediate reaction was, as to who will read it? In any case, there is no dearth of serving experts with superior domain knowledge.

After some deep reflection and search, it emerged that there is no stand-alone book on Infantry, Artillery or Army Air Defence which one can refer to in a synergized and holistic manner to acquire awareness on the subject. No doubt, there are numerous books on military history, strategy, principles of war and coffee table books but if I want to pick a book on the aforementioned subject then the chances are that I may not succeed. The other factor which came to my mind was that domain knowledge matured with four decades of experience will certainly contribute to a better understanding of the subject, especially useful for educating young minds who are just starting their career. Such books can also help the civil society to learn more about the defence forces so that they can give an informed and a knowledgeable response when required.

Coming to this book proper on 'The Future Needs of GBAD in the Country', I can say that it is a labour of love where VK has not only written several chapters himself, but also, has requested various distinguished experts to contribute chapters on subjects of their respective expertise. This has resulted in a rich tapestry of interwoven knowledge which traces two parallel narratives of the evolution of 'air power' and its counter force of 'GBAD'. The chapters cover all aspects of GBAD, from 'Air Threat' which is the raison d'être for GBAD, to Basic Concepts, to Battle Management Systems and ending with Ballistic Missile Defence which, to my mind, will be the final frontier of GBAD. It will be proper to state that advanced countries like the United States of America and others relied almost solely on air power for air defence and thus, there was scanty development of GBAD weapons. The weapons of the Sixties were Hawk and Chaparral missiles. Sergeant York was developed as a divisional AD self propelled gun system but was shelved later on. The Russians on the other hand, produced a plethora of GBAD weapons like Schilka, Kvadrat, OSA-AK and so on. The Americans have realized the importance of GBAD weapons but they are relying solely on missile systems like Patriot and others. May be this book will leverage them to also focus more deeply on ground based systems for Air Defence.

I would go far as to say that this book is a treatise or a Bible on GBAD and will serve well to all discerning readers who want to refer to it. I must compliment VK for his sterling endeavour and wish him all the best for his future projects.

PREFACE

Why this Book?

Before I put pen to paper, the first thought which came to the mind was— Why this book? Why one more addition to the existing body of written work? The answer seemed to echo from the multiple voices I hear from the environment...

Why are so many different types of guns and missiles required for air defence? Why can't one missile be a standard fit? Why different ranges and altitudes cannot be achieved by changing the deployment locations and pattern of a few standard weapon systems?

Enemy air threat is no issue; we have such a strong Air Force that it will not let the adversary move. We will actually require no ground based air defence as our Air Force will wipe out the enemy's air force from the skies within 48 to 72 hours of the balloon going up.

Ground Based Air Defence Weapon Systems (GBADWS) are extremely costly; if all the funds are diverted towards them, little will be left for others.

None of the above statements and queries are flights of fancy. These are the arguments actually transacted between stakeholders and decision-makers at various points of time.

In addition to the above, it has been my personal experience formed over decades as an air defence professional that while there is a large volume of written word and reference documents available on other frontline combat arms such as infantry, armour, artillery, signals and more, there is a general lack of it when it comes to the sharp vertical of ground based air defence (GBAD) including ballistic missile defence (BMD) as a part of the overall aerospace superset.

Thinking of GBAD, some very basic and bottom line questions come to mind which demand a sound answer. The questions run like this...

- What are the contours of the air threat we face today from our potential adversaries? How are these going to metamorphose in the foreseeable future, given the ongoing feverish pace of activity in the adversary camp aimed at revamping their 'air strike punch'?
- What is the overall generic requirement of GBADWS to counter the appreciated air threat?
- Do we have enough (as gleaned from open sources)? Which way must our future development and procurement proceed so as to provide a viable answer to the current and futuristic air threat, both in qualitative, as well as, quantitative terms?
- How will we fight the 'emerging technologies' in the aviation and aerospace domain with counter technologies in the GBAD sphere?

It is for addressing these questions and more and for providing to a professional reader, a researched and an analysed answer to the above major issues, that I have endeavoured to make an attempt to bring out this book.

In living up to the above mandate, the book also aims to serve the purpose of filling a void that exists in the domain of well-researched and authentic reference material related to the vertical of GBAD.

I hope this book serves the purpose for which it has been written.

LIST OF CONTRIBUTORS

Lt Gen (Dr) V K Saxena, (Retd), PVSM, AVSM, VSM

Gen Saxena is a former Director General of the Corps of Army Air Defence. He is a Distinguished Fellow at the Vivekananda International Foundation and holds the Kothari (DRDO) Chair of Excellence at the United Services Institution of India. He is also a United Nations Scholar with an M.Phil. and Ph.D. and a law scholar pursuing qualifications in Human Rights, Child Rights and Medical Law and Ethics at the National Law School of India University (NLSIU),

Gen Saxena has been decorated three times by the President of India. Besides, he is a recipient of numerous prestigious awards in the Service. He has also been awarded by the Institution of Electrical and Telecommunication Engineers (IETE) for his seminal contribution in the development of Low Level Light Weight Radar by the DRDO.

Gen Saxena is a prolific writer. He has authored five books and has to his credit more than 115 articles in various defence magazines covering the subjects in his core competency domain of air defence, aerospace, military communication, cyber security, nuclear security and defence procurements.

His earlier books include Ground Based Air Defence - Contemporary Issues, Amazing Growth Journey of UAVs and Ballistic Missile Defence Capabilities, Future of United Nations in the 21st Century and United Nations and Human Rights - Past Performance and Future Challenges

He also spearheads a project for the welfare of families of the forces casualties and disabled soldiers.

Lt Gen Naresh Chand, (Retd), PVSM

Gen Chand is the former Director General of the Corps of Army Air Defence. He has also been the Commandant of the prestigious Army Air Defence College in its formative years. He has written hundreds of articles across the entire spectrum of defence magazines and journals. He has also produced coffee table books and regimental histories. His subject range has not only been air defence but also the same resides in the domain of all the three Services. He has served with DRDO in their Electronics and Radar Development Establishment. He has also been member-secretary of the erstwhile National Radar Council which was then chaired by Professor M.G.K. Menon, who played a prominent role in the development of science and technology in India.

After hanging up his boots, he was a consultant with Bharat Electronics Limited for the upgrade of Schilka, 40mm L/70 and 23mm twin gun. All these systems are currently being inducted into service. At present he is serving as the Senior Technical Group Editor at the SP's Group of Publications which publishes defence-related magazines and books.

The officer enjoys the position of a valued scholar and a highly respected air defence warrior.

Air Marshal Anil Chopra, (Retd), PVSM, AVSM, VM, VSM

Air Marshal Chopra is a Subject Matter Expert of great repute and standing on all matters of air, air defence and aerospace. He is a respected and a prolific writer who has the credit of getting published all across the spectrum of defence magazines and journals month after month. He is a highly decorated test pilot, a pioneer of the Mirage-2000 fleet who has had the honour to command a Mirage squadron and the prestigious Aircraft and Systems Testing Establishment (ASTE) of the IAF. He also commanded operational airbases in both the Western and Eastern sectors.

He has been at the helm of the IAF's Operational Inspection Directorate (Directorate of Air Staff Inspections or DASI) and was Air Officer Commanding Jammu & Kashmir. He retired as the HR head of the IAF. He has been a member of the prestigious Armed Forces Tribunal. He was a member of the Executive Council of Jawaharlal Nehru University (JNU) for two years. In addition to four Presidential awards, he is the recipient of the Gandhi Peace Medal for his work in J&K.

Relevant to this chapter, the Air Marshal has headed the tri-service team that evolved the very concept of ASC.

Dr V K Saraswat, Member Niti Aayog

Dr Saraswat is a national icon of great standing who is currently serving as a member of NITI Aayog. He has been the former Director General of the Defence Research and Development Organisation (DRDO) and the Chief Scientific Advisor to the Hon'ble Raksha Mantri.

He is Chancellor of Jawaharlal Nehru University and the recipient of the Padma Shri and Padma Bhushan from the Government of India. Besides, a host of other awards such as Vikram Sarabhai Memorial Award from Indian Science Congress (2011); Prof Jai Krishna Memorial Award by Indian National Academy of Engineering (INAE); Dr. Y. Nayudamma Memorial Gold Medal for the year 2011 by AP Science Congress; Jawaharlal Nehru S&T Award (2009) by Government of Madhya Pradesh; Arya Bhatta Award (2011) from Astronautical Society of India and National Aeronautical Prize (1998) have been conferred upon him. Dr. Saraswat has also been conferred honorary doctorates from more than 25 Indian universities.

Dr Saraswat is considered to be the father of the Prithvi missile system, as well as, the Indian Ballistic Missile Defence programme, codenamed Programme AD. He has worked tirelessly and has guided thousands of experts in steering these programmes from their inception, right up to their current state of maturity.

Brig GS Choudhry, VSM

Brig Choudhry is an officer with a tremendous track record. Having done all courses of repute and prestige that are applicable to his rank both in India and abroad, the officer has earned the unique distinction of being a subject matter expert on air defence and aerospace. Having been the only officer selected from his batch to the General Cadre, he is new commanding a field formation with great success.

The officer has been a prolific writer whose works keep getting published in a host of defence magazines and journals. He has a penchant for emerging technologies and takes great interest in keeping himself abreast with the latest. A sports enthusiast and a marathon runner, the officer is contributing immensely in the field of air defence and aerospace besides a host of other subjects.

Lt Gen Ram Pratap, (Retd) , PVSM, AVSM, VSM

Gen Ram Pratap has been the former Director General of the Corps of Army Air Defence. A graduate of the Canadian Command and Staff course, Toronto and Kvadrat Weapon System in Mari (USSR), he had the honour to command the prestigious Army Air Defence College, as well as, Army Air Defence Centre and thus oversee the training of a whole generation of air defence warriors.

Gen Ram Pratap is a conceptual thinker whose sharp intellect and the capability to generate new ideas are respected across the entire environment. He is a subject matter expert of great repute and standing in matters related to air, air defence and aerospace. He has not only written extensively on the above subjects but also, has put into place many practical and innovative ideas which have helped a great deal in sustenance and upkeep of the vintage inventory of ground based air defence weapon systems. The officer has done seminal work in various conceptual studies and research work besides heading the all-important Cadre Review of Soldiers of the Army.

Brig MKK Iyer, (Retd) SM

Brig Iyer is an expert on air defence and aerospace matters. Besides this, he has decades of hands-on experience in all matters related to defence procurement. He commands that level of knowledge, respect and acceptability in the professional environment which only a handful of experts do.

The officer is an alumnus of the Higher Command course and has held many a prestigious appointment in the Army. He is currently been selected to serve by the CII to be on board the Society of Indian Defence

Manufacturers (SIDM) where he is contributing a great deal in progressing the cause of Make in India. The author has written extensively on many subjects related to the conceptual, technical and procedural aspects of air defence, aerospace and defence procurement.

He has had first-hand experience of handling simulators in the Army and has been at the forefront of generating new and innovative ideas related to the use of simulators in the Services.

LIST OF ABBREVIATIONS

AAR	Air-to-Air Refuling
AC	Aircraft (s)
AD	Air Defence
ADBMC2	Air Defence Battle Management Command and Control
ADC&RS	Air Defence Control and Reporting System
ADGM (SP)	Air Defence Gun Missile (Self Propelled)
AEW&C	Airborne Early Warning and Control
AFNET	Air Force Net
AH	Attack Helicopters
AHEAD	Advance Hit Efficiency Air Defence
ARM	Anti Radiation Missiles
ASMs	Air to Surface Missiles
AVIC	Aviation Industry Corporation of China
AW	All Weather
BDA	Battle Damage Assessment
BMC2	Battle Management Command and Control
BMD	Ballistic Missile Defence.
CAO	Counter air Operations
CIWS	Close in Weapon Systems
CRV	Class-Room Variants

COMINT	Communication Intelligence
CPB	Charged Particle Beams (CPBs)
CSFO	Counter Surface Force Operations
CKP	Commutative Kill Probability
DPSU	Defence Public Sector Undertaking
DTTI	Defence Technology and Trade Initiative
ed.	Added by the editor
ESM	Electronic Support Measures
EW	Electronic Warfare
FAA	Federal aviation Authority
FLIR	Forward Looking Infra Red
GA	General Aviation
GBADWS	Ground Based Air Defence Weapon Systems
GBAD	Ground Based Air Defence.
GPS	Global Positioning System.
GCS	Ground Control Station
HEPTRS	Helicopters
HMI	Human-Machine Interface
HPM	High Power Microwaves
ICAO	International Civil Aviation Organisation
IDS	Integrated Defence Staff
IFF	Identification Friend or Foe
INS	Inertial Navigation System
IP	Internet Protocol
IR	Infra Red
ISTAR	Intelligence, Surveillance, Target Acquisition and Reconnaissance

LGB	Laser Guided Bombs
LRSAM	Long Range Surface to Air Missile
LTA	Lighter Than Air
LTIPP	Long Term Integrated Perspective Plan
MOPs	Mobile Observation Posts
MTI	Moving Target Identification
MUM-T	Manned And Unmanned Teaming
MRs	Military Regions
MRSAM	Medium Range Surface to Air Missile
NOE	Nap of the Earth
NAS	National Air Space
PAF	Pakistan Air Force
PGMs	Precision Guided Munitions
PLAAF	Peoples Liberation Army Air Force
PPP	Public Private Partnership
QRSAM	Quick Reaction Surface to Air Missile
RASP	Recognised Air Situation Picture
RCS	Radar Cross Section
ROI	Radii-of-Action
RSTA	Reconnaissance Surveillance and Target Acquisition
SATCOM	Satellite Communication
SEAD	Suppression of Enemy's Air Defence
SIGINT	Signal Intelligence
SPM	Strategic Partnership Model
SSGs	Signal Scenario Generators
SSMs	Surface to Surface Missiles.
SRSAM	Short Range Surface to Air Missile

STOVL	Short Take off and Vertical Landing
TBM	Tactical Ballistic Missile
TMC	Tibet Military Command
TSTs	Transportable Satellite Terminals
UAOP	Unmanned Aircraft Operator Permit
UAS/UAV	Unmanned Aerial System/Vehicles
UIN	Unique Identification Number
VSHORADS	Very Short Range Air Defence System
WAN	Wide Area Network
WMD	Weapons of Mass Destruction
WTC	Western Theatre Command

CHAPTER 1

The Changing Contours of Air Threat

Lt Gen (Dr) V K Saxena, (Retd), PVSM, AVSM, VSM

Air power may either end war or end civilisation[1] – Winston Churchill, House of Commons, 1933

Since the raison-de-etre for the existence of air defence on land, sea or air is to counter the air threat from an adversary, an assessment of the same, both in terms of quality and quantum, must therefore become the ab-initio of all deliberations concerning air defence.

The Cause-Effect Relationship

The term 'air threat' essentially implies the cumulative potential danger which our adversary holds, as his capability to cause damage and destruction to our vulnerabilities through the use of aerial threat vehicles and munitions in furtherance of, or in support of his war effort. This danger is basically assessed in two dimensions. The first is the qualitative dimension, which includes the lethality, severity and technological advancement of the adversary's air threat vehicles, while the second is the quantitative dimension that relates to the numbers which constitute the air threat punch. Accordingly, the defender aims to put together such resources on land, sea and air that can counter the threat as defined above. Thus air threat on one side and air defence on the other are tied to each other in an eternal cause-effect relationship.

The Two Verticals of Metamorphosis

The multi-dimensional growth of air threat over time, has mainly taken place along two major verticals. The first of these is the multiplicity of air threat vehicles. In that, the erstwhile duo of aircrafts (ac) and helicopters (heptrs) as main threat vehicles, are today joined by a multi-dimensional strike punch consisting of attack helicopters (AHs), unmanned aerial systems/vehicles (UAS/UAVs), tactical ballistic missiles, cruise missiles, precision guided munitions (PGMs), anti-radiation missiles (ARMs) and the now emerging soft-kill offensive weapons.

The other vertical is the spatial growth of the threat in all its three dimensions of range, reach (implying altitude) and hours of applicability. Starting from say, the early Seventies when the air threat was prosecuted mainly in the visual domain of the defenders by mostly using short-range guns and a small arsenal of unguided rockets and missiles, today, the capability is to strike deep with precision and accuracy from long ranges by standing off hundreds of away from the defenders' visual domain. There is also the capability now to strike with multiple threat vehicles at any time of the day or night, remain nearly invisible to defenders' radars, employ UAS with their multiple combat virtues and hit both with hard and soft kill munitions.

Manifestation of the Metamorphosis in Threat Vehicles

The revamping-transformation of the air threat over time, as stated above, is spread over the entire spectrum of multiple threat vehicles as enumerated above. Today, these fourth and fifth generation war machines are truly multi-role, structurally lighter and detection resistant. These carry an array of on-board sensors with a packaged electronic warfare (EW) suit for survivability and sustainability and are armed to the teeth with smart and intelligent munitions including the PGMs. The AHs have become all-weather (AW) with avionics providing the technological muscle for nap of earth (NOE) operations. The unmanned revolution has taken over with its unmatched capability of all-weather, all-terrain operations with incredible endurance and no crew fatigue. Multiplicative affects are being achieved by manned and unmanned teaming (MUM-T). The missiles, both in the surface-to-surface (SSM) and air-to-surface (ASMs) modes are scaling new dimensions in range, reach and kill probabilities. Soft kill weapons are emerging in three domains, viz., laser, charged particle beams (CPBs) and high-power microwaves (HPM).

Against this background, a brief analysis of air threat from Pakistan and China follows.

Pakistan

The mission of the Pakistan Air Force (PAF) is to provide, in synergy with the other Services, the most efficient, assured and cost-effective aerial defence of Pakistan.[2] The PAF has an inventory of some 450 combat aircraft organised into 20 front-line combat squadrons and over 300 support aircraft. Its main types of combat aircraft include the French Mirage V strike aircraft (90), Mirage III interceptors (90), Chinese J-7 fighters (39), JF 17 multi-role aircraft (86) and the US F-16 multi-role strike aircraft (76). It is reported that by 2025, the PAF will operate mainly two multi-role aircraft, namely, F-16 and JF-17 and FC 20.[3]

The Future Face of Pakistan Air Force - F-16, JF 17, FC 20

Over a period of time, old and obsolete aircraft (A5, F6, Super F7, FT7P) have been phased out. The F-16s have undergone major upgrade to include NATO compatible Link-16 communications, global positioning system/inertial navigation system (GPS/INS)-aided flying, weapon upgrades to include high speed ARMs, PGMs, stand-off weapons and wind-corrected munitions dispensers. The JF-17 has been operationalised with high-grade weaponry, night-flying capability and EW package. Similarly, Mirages have been put through systems and weapon pack upgrades.

Talking of AHs, the PAF has a fleet of AH 1F Huey Cobras (2 squadrons, 32 heptrs) and the Chinese K-8 (1 squadron, 28 heptrs). The AH have seen upgrades involving forward looking infra red (FLIR) night sights and TOW missiles. Besides the AHs, the PAF operates a small fleet of UAS (covered later). In September 2015, Pakistan became the ninth nation in the world to develop and use an armed UAS named Burraq. It is armed with Barq, an ASM with pin-point precision.

AH 1F Huey Cobra

Under the Armed Forces Development Programme (AFDP 2006-2019), the PAF, besides procuring modern combat aircraft from the US and China, has also acquired several cutting-edge systems. These include PGMs for J-17 aircraft (LS6 and LS6P and SD10A Beyond Visual Range Missiles). Other known PGMs and ASMs with the PAF are Shrike ARM, laser-guided bombs (LGBs), Maverick missiles and AS 30 ASM useable with Mirages (III&V) and F 16 aircraft. As regards the cruise missile, Nasr (HATF IX) is a solid-fuelled tactical ballistic missile (TBM) developed as a short-range (60 km) high accuracy, low yield battlefield deterrent.

On 19 September 2017, the PAF announced that it will acquire three new Saab 2000 Erieye airborne early warning and control (AEW&C) aircraft. This will join the three existing Erieyes and four Y8600-based ZDK03 AEW&Cs, taking it to a total of such aircraft to 10 by the end of the decade.[4] The first of these aircraft was likely to be delivered by December 2017 and the remaining two by December 2018. This will provide a tangible enhancement in the airborne surveillance and targeting capability of the PAF. It has also been reported that Pakistan is in talks with China to acquire 30-40 JF-31 Stealth fighters and with Russia for Sukhoi Su-35 air-superiority multi-role fighters. The same will add a totally new qualitative dimension to its strike capability.

In the unmanned category, while the Burrack represents the front edge, UAVs from China (ASN 105A, ASN 206, B2B,) UK (Snipe Mk II), Germany (Luna), Italy (Falcon) and South Africa (Seeker) besides the indigenous (Bravo, Ababeel, Baaz, AWC Mk I & II, Vision I, Shaper, Vector, Huma 1 and Thunder) constitute the total UAS fleet.

Pakistan has also had a very successful run of their surface-to-surface missiles (SSMs) in providing delivery means for their conventional, as well as, nuclear warheads (assessed 120-130 but of smaller yield and reaches than India[5]).

An attempt is made to trace the evolution of the Pakistani Missile Programme.

The Trigger—Indian Missile Development Programme

In 1982-83, the Indian Ministry of Defence (MoD) launched the Integrated Guided Missile Development Programme (IGDMP) to be managed by the Defence Research and Development Organisation (DRDO). The IGMDP was designed to develop four different types of missiles, namely, short-range surface-to-surface missiles (SSM) code-named Prithvi, short-range low level surface-to-air missile (SRSAM) code-named Trishul, medium-range SAM (MRSAM) code -named Akash and third generation anti-tank guided missiles (ATGM) codenamed Nag. The Agni series of missiles were initially conceived as technology demonstrators; these was later upgraded to ballistic missiles with different ranges (1100-5000+Km) under the IGMDP banner.[6]

The felt need for the SSMs was to provide a delivery means to both conventional and nuclear weapons deep into the enemy's hinterland while the SAMs were to provide the capability to thwart the growing severity and lethality of the air threat posed by our potential adversaries. The anti-tank missiles were to create a capability against enemy armour.

The Response from Pakistan

In response to India's planned build-up of SSM and SAM capability, Pakistan's missile programme began around 1987. The initial motivation for this programme was to counter the 'missile gap' created by India's testing of the Prithvi missile system and was aimed to develop short to medium range missile systems to deter missile threats from India.

The missile programme adopted an umbrella code-name—HATF meaning "Vengeance." In Arabic, HATF means "death" and refers to the sword of Muhammad which was used in many of his military conquests and had the unique distinction of never missing its targets. Looking from Pakistan's point of view, it was an apt name to a series of missiles that it intended to use against India (in vengeance of course!)

The Ballistic missiles worldwide are classified as short, medium, intermediate and intercontinental range missiles, abbreviated as SRBM, MRBM, IRBM and ICBM respectively. The range brackets are: SRBM 300 -1000 km, MRBM 1000-3500 Km, IRBM- 3500-5500 km and ICBM >5500 Km.

Under the above Programme a series of SSMs were developed. These include Hatf I-SRBM, range 80–km, payload 500 kg, status—operational. Hatf II (Abdali) named after the eighteenth century Afghan conqueror, SRBM, range 200 km, payload 500 kg, status—operational. Hatf III (Ghaznavi), named after the invader Mohd. Ghaznavi, SRBM, range 290 km, payload 500 kg, status—operational. Hatf IV (Shaheen 1), Shaheen meaning death, SRBM, range 750-900 km, payload 850 kg, status—operational. Hatf V (Ghauri I and II): Ghauri 1 MRBM, range 1500 km, payload 750 kg, status—operational, Ghauri II MRBM, range 1800 km, payload 750-1200 kg, status—operational. Hatf VI (Shaheen II) MRBM, range 2500 km, payload 700 kg, status—operational. Shaheen III, range 2750 km, payload 700 kg, status—operational. Hatf VII (Babur), land attack cruise missile, range 700 km, payload 500 kg, status—operational. Hatf VIII (Ra'ad), air launched cruise missile or ALCM, range 350 km, payload 500 kg, status—operational, Hatf IX (Nasr), quick-reaction shoot-and-scoot missile ideal for launching tactical nuclear weapons or TNWs, status—operational.

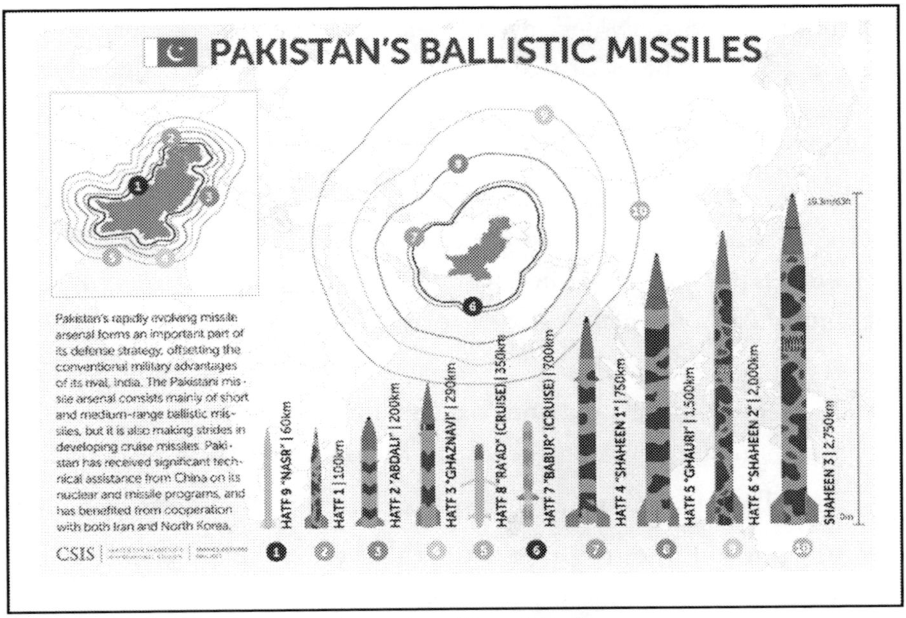

Pakistan's Ballistic Missiles

Among the ballistic missiles mentioned above, two types demand special mention. The first are the cruise missiles that reside on the cutting edge of technology and bring in a measure of high speed (Babur 1 - 880 km/h) and high accuracy. Babur II launched on 14 April 2018 reportedly destroyed its intended target "within metres." The second type of SSM in the Pakistani arsenal deserving separate mention is Nasr (Hatf IX). These are quick-reaction shoot-and-scoot missiles Nasr is an SRBM with a range of 60 km. It is a multiple launch rocket system capable of carrying four ready-to-fire nuclear-capable missiles with a weapon yield in the region of 0.5 to 5 KT. Starting from its first flight test on 19 April 2011, the missile was quickly claimed to be operational on 5 October 2013. As is evident from the weapon yield mentioned above, missiles like Nasr are suitable to launch TNWs that bring in a totally new dimension to the entire spectrum of nuclear deterrent (not discussed further).

As for TNWs, it is stated that in order to play the nuclear blackmail game with India, (but keep it below the threshold of attracting punitive nuclear retaliation) and under its garb, conduct nasty acts of terrorism, Pakistan has reportedly developed lower yield nuclear weapons (TNWs) that are capable of being carried accurately by Nasr tactical ballistic missile having multiple tubes for launching several TNWs together (test fired on 5 July 1917).[7]

In the larger context, the spectrum of ballistic missiles developed by Pakistan basically provides it with the capability to deliver nuclear warheads on its adversary. The above said capability is considered 'realised' only when the three pillars on which it stands are realised. These pillars are:

1. Nuclear warheads.
2. Delivery means on land, sea and air.
3. Command and control aspects including field infrastructure.

The following points are stated in the context of the above three pillars:

An open-source estimate puts the Pakistani nuclear stockpile at 110-130 warheads with a capability to make 20 devices annually[8] (not discussed further).

While ground-based delivery means in terms of SSMs stand described above, a careful examination of the details presented would reveal that as a ground delivery system, Pakistan has a family of eight different types of ballistic missiles that could serve as ground-based delivery systems for its nuclear arsenal.[9]

As regards sea-based delivery capability, Pakistan on 1 April 2018, reportedly tested a nuclear-capable submarine-launched cruise missile (SLCM), Babur II, the land attack version of which was tested on 14 April 2018. It is suspected to be fitted on the French made diesel-electric submarine Agosta 90 B—Khalid Class, or is likely to be fitted in future on the S-20 Yuan class submarines that Pakistan is likely to acquire.[10] The SLCM Babur II has a range of 450 km.[11]

As regards the air delivery capability, it is reported that Pakistan has three types of aircraft that can be used to deliver nuclear weapons. The first is the F-16 that can be equipped to toss-bomb nuclear warheads; the second is the Mirage III/V and the third is the JF 17 S. The last two could be equipped with Hatf VIII air launched cruise missiles (ALCM) capable of carrying nuclear warhead.[12]

As regards the third pillar of strategic capability that is the command and control aspects, the following points are made in brief.

The National Command Authority (NCA) was established in 2000. It is the apex civilian-led (the PM of Pakistan is its chairman) command to oversee the employment, policy formulation, exercises, deployment, research and development and operational command and control of Pakistan's nuclear arsenals.[13] It comprises two committees, namely, the Economic Coordination Committee (ECC) and the Development Control Committee (DCC) besides a Strategic Plans Division (SPD).

The ECC is responsible for directing policy-making during peacetime, making recommendations for evolution of nuclear doctrine, establishing the hierarchy of command and the policy for authorising the use of nuclear weapons and establishing guidelines for an effective command and control system to safeguard against accidental and unauthorised use.

The DCC is responsible for exercising technical, financial and administrative control over the strategic organisations involved in the nuclear weapons programme and overseeing the development of programmes for the strategic forces.

The SPD is responsible for formulating policy options (nuclear policy, strategy and doctrine) for the NCA. It is also responsible for implementing the NCA's decisions and drafting strategic and operational plans for the deployment of strategic forces.[14]

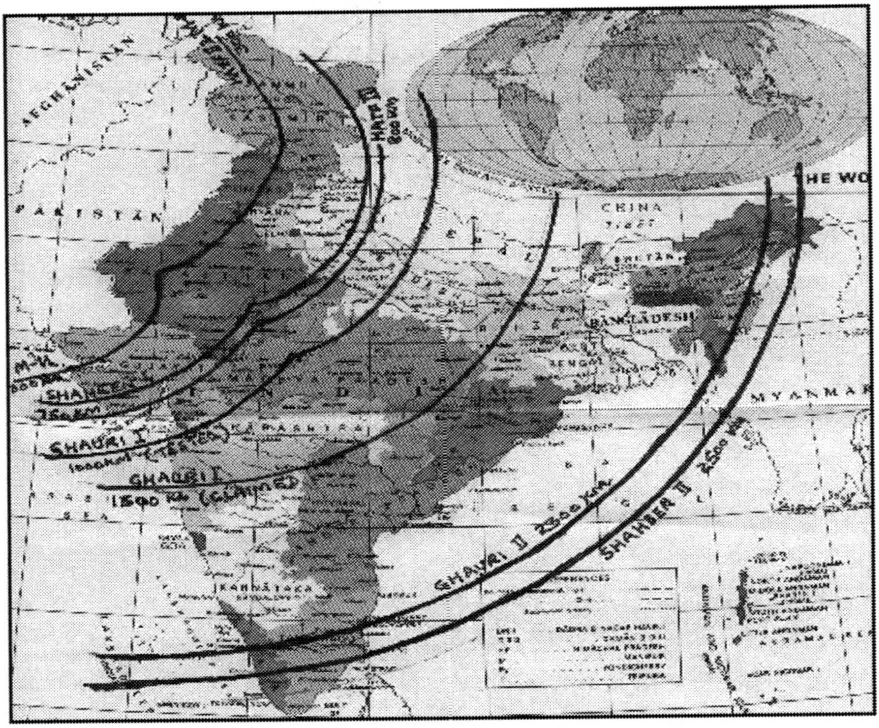

Reach of Pakistani Ballistic Missiles into India

A look at the range and reach capability of Pakistan's ballistic missiles reveals that it has the capability to reach well inside India's hinterland and beyond. It also clear that the missiles have the capability of delivering both conventional as well as nuclear warheads. The triad of delivery means as enumerated above, also indicates that Pakistan has options of delivery from all the three mediums of land, sea and air.

Talking of the latest in the domain of Pakistani ballistic missiles, it was reported by the Inter-Services Public Relations (ISPR), the media wing of the Pakistan Military, that on 14 April 2018, Pakistan had successfully test-fired the enhanced range version of its indigenously developed cruise missile named Babur which is capable of striking targets on land and at sea.

Babur 1(B) was launched from a land-based transporter-erector-launcher (TEL). Though the missile is reported to have a range of 700 km, the report did not specify, up to what range it was tested. It was claimed that the missile is equipped with niche technologies in navigation called terrain contour matching (TERCOM) and all-time digital scene-matching area co-relators (DSMAC). Both these combined, enable the weapon system to engage various types of targets with pinpoint accuracy even in the absence of GPS navigation. The missile can carry both conventional, as well as, nuclear warheads.[15]

Launch of Babur I(B) 14 Apr 2018

How Pakistani Ballistic Missile Capability Needs to be Addressed?

While the own SSM programme as enumerated earlier continues unabated in an eternal cause-effect cycle, SSMs do not defeat SSMs. It is the anti-missile capability (referred to with the generic name of Ballistic Missile Defence or BMD) that defeats an adversary's SSMs.

In the above context, it is known that the indigenous capability in the BMD domain is being developed by the DRDO under its programme, code-named 'Programme AD'.

The details about this programme have been covered in Chapter 6.

Another important capability that needs to be mentioned in the context of the ballistic missiles of the adversary is the capability to launch multiple independently targetable re-entry vehicles (MIRVs) by ballistic missiles that have the capability to fly outside the Earth's atmosphere (exo-atmospheric) for most of its flight and launch MIRVs upon re-entry into the Earth's atmosphere.

While India is still some distance away from MIRV capability, Pakistan seems to be closer to it. It is reported that Pakistan has been developing a missile called Ababeel (range 2200 km) which is said to be capable of carrying an MIRV warhead.[16]

In this context, a statement by the Chinese Academy of Sciences (CAS) reported in the *South China Morning Post* (SCPM) on 22 March 2018 said that China has provided Pakistan a highly advanced optical tracking and measurement system which will be a boon for its missile programme

This system is capable of taking precision images of a missile trajectory from launch till the end. The image trail, which is on a continuous thread, is of extremely high resolution and lends itself to presentation on digital media in a variety of picture formats in real time. Also, since the entire thread is digitally recorded, it is preserved for review later by analysts and provides such common features as rewind, fast forward, pause, zoom in/out and more.

Chinese advanced tracking system sold to Pakistan

The sale of the advanced tracking system could have the following impact:

1. It will greatly help Pakistan in testing and perfecting its ballistic missile arsenal for higher ranges, reaches and accuracies. Of

particular relevance is Ababeel whose operationalisation may be hastened with the induction of the said tracking system. Talking of Ababeel, the logic of four telescopes in the EOTS also falls in place considering the requirement of independently tracking, measuring and analysing MIRV payloads and their independent trajectories.

2. The induction of this capability will give a quantum jump to Pakistan in its capability to test and perfect SSMs in general.
3. It is a step by China to continuously ratchet up the SSM threat to India. This time through its 'all-weather friend' by opening another window of threat from the western front.
4. In an unprecedented and rare gesture of, firstly, going through with the sale of such a sensitive equipment to a foreign buyer and, secondly, making it public, China has attempted to send a strategic message to India close on the heels of the launch of Agni V (18 January 2018).

As regards the implications for India, the following is stated:

1. As stated, SSMs do not counter an adversary's SSMs; it the ballistic missile defence (BMD) capability that counters an adversary's ballistic missile threat. Therefore, it becomes imperative for India to hasten up the operationalisation of its BMD capability in 'Programme AD'. The following specific actions are required:

The BMD system must now move from scientists to users and become a part of Strategic Forces Command (SFC) as deployable fire units in the earliest time frame.

In order to make the BMD capability 'deliverable,' there is a need to build robust and survivable infrastructures, a fail-proof and survivable regime of codes and authorisations for weapon delivery duly supported by several layers of redundant and secure communications.

Keeping the compulsions of 'No First Use' (NFU), the survivability of actors post the first strike needs also to be ensured through alternate command posts with a chain of hierarchy fully defined and known to the decision makers up and down the chain.

Also, not merely as a mindless rat race, but as a logical build-up of capability riding on the wheels of progression of technology, our MIRV capability on board Agni VI must also be pushed to fructification.

Some other implications from the PAF capability that need be factored into our quality and quantum calculation of air defence resources in a tri - service format is covered below:

- 20 combat squadrons of PAF comprise 17 strike, 4 multirole and 3 air defence aircraft squadrons.
- There has been a continuous qualitative upgradation wherein the dead weight of old and vintage aircraft is being done away with, sharp-focussing the fleet to four major front-line strike aircraft types (F-6 series, Mirage III/V, J7/JF17).
- There has been a continuous attempt to re-vamp the 'threat beyond aircraft' in terms of AH, cruise missiles, ARMs, SSMs, UAS, and PGMs.
- A limited night/AW attack capability and a tangible mix of EW capability stands inducted in the front-line aircraft.
- Limited induction of AWACs and air-to-air (AAR) capability provides the force multiplier effect.
- The front edge of the strike force now features the stealth threat which needs to be addressed by the defenders.
- The known radii-of-action (ROI) of the front-line combat aircraft provides the capability to the PAF to strike up to India's hinterland making the strategic assets in our rear areas vulnerable

China

In keeping with China's high growth trajectory, the People's Liberation Army Air Force (PLAAF) has followed an impressive growth profile. From a near-antiquated status around 1990-91, the PLAAF today is more operationally able today than at any time in its past.

As part of a major revamp of its air power, its earlier inventory of some 5000+ ac that included some 2000 obsolete aircraft stands right-sized to about 3000 ac (3010[17]) in which, the combat edge is held by J-7 (fighter-388), JH-7 (fighter bomber-70), H6 (strategic bomber-120), Su 30 (air superiority fighter 76), J-8 (Interceptor 96), J-10 (multi-role fighter-237), J-11 (air superiority fighter-196) and Q5 (strike aircraft-118) ac.

Taking pride in flying the first prototype of its latest stealth fighter (J-20) in January 2011 ostentatiously at a time when the US Secretary of Defence, Mr. Robert Gates was in Beijing, the PLAAF is steadily modernising. Predictions on the future PLAAF fleet indicate that it will consist of large numbers of Chengdu J-10 and Shenyang J-11 as mainstay platforms and JH-7A as its backbone precision strike fighter, J-20, is likely to enter PLAAF around 2018.[18, 19]

J 10

J 11

JH 7A

J 20

Mainstay aircraft of the Chinese Air Force

The mainframe of its future transport fleet is likely to be anchored around Y-9 medium range transport aircraft duly supported by its IL fleet. The AH inventory is likely to be populated by WZ-10, WZ-9 and Z-11. The AWACS fleet is likely to have refined variants of KJ2000 and KJ200. Also, a reported 50, (unconfirmed) number of similar platforms are likely to be imported from Russia. The Aviation Industry Corporation of China (AVIC) is reported to be working on six developmental combat aircraft programmes. These include the significantly enhanced Chengdu J-10B multirole fighter, Shenyang J-11B5 heavyweight strike fighter, Shenyang

J-15 carrier-borne fighter and Chengdu J-20 fifth generation fighter demonstrator. There is at least one more fifth-generation fighter project yet to be publically reported. Besides this, the continuing enhancement of current frontline types such as the JH 7A and H-6 bomber family is in progress.[20]

The PLAAF is aiming to be one of the world's foremost air forces by 2020, with at least 1,000 'modern' combat aircraft. The Russian Su-35 aircraft, along with its advanced IRBIS-E passive electronically scanned array radar system is likely to enter service by 2018 and will significantly enhance the PLAAF's ability. Of greater concern also is the offensive capability in terms of PGMs and surface-to-surface missiles.

The PLAAF has a very strong muscle in UAVs. Besides the existing inventory of UAVs (Xianglong, WZ-9, Yi Long, CH-3, Anjian, etc.) some additional inputs are as under:

> Chinese UAV experiments have shown impressive levels of innovation by building and flying unmanned vehicles of every conceivable design and genre. These include stealthy UAVs, morphing UAVs, annular wing vertical takeoff and landing (VTOL) UAVs, micro-UAVs, unmanned airships, flying wings, modified light aircraft and sailplanes, UFO-style flying discs and even orinthopters (UAVs utilising flapping wing design). Innovators, both professionals as well as greenhorns/debutants, have been allowed to run free and explore almost every UAV configuration.
>
> Two other features are clearly visible in Chinese UAV development, namely, arming of UAVs and equipping them with enabling sensor payloads. China has followed a logical process in arming its UAVs by first adapting existing air-to-surface weapons and then developing purpose-built small light weight munitions. In each case, the weapons involved are relatively new. Weapon-vehicle matching has been thoughtful. Some examples are Wing-Loong and CASIC UAV.

Besides the conventional payloads, the PLAAF is emphatically present in the field of smart, intelligent and precision ammunition with surgical strike capability. Way back in 2007, China developed a powered smart bomb (KD-88) having an infra red (IR) TV guidance system. The estimated range is about 110 km with a capability to hit small targets.[21] The other state-of-the art arsenal includes the supersonic Russian and Chinese made ARMs (KH 31P, YJ-91) for operational use with SU-30, JH 7A and J-11 ac and LASER- guided and satellite-guided bombs (on bd Q-5 ac).[22] It is now estimated that almost all the modern aircraft of the PLAAF today are capable of carrying the PGMs.[23]

Chinese Missile Development Programme

The history of Chinese missile development dates back to its efforts to make its first nuclear bomb. According to Xinhua sources the great decision was taken in 1955. The erstwhile USSR provided the Chinese with some SS2 'Sibling' short range ballistic missiles in the mid 1950s. The Chinese built a small number of missiles called 'Dong Feng' (East Wind)-1. The US designation of these missiles runs with a prefix of CSS or SS. The first DF-1 was launched in 1960. Thereafter, the Chinese successfully launched guided missiles in October 1963.

Presently, China has a well-developed ballistic missile programme to support its minimum nuclear deterrence policy. Chinese ballistic missiles are seen as both conventional and nuclear delivery systems. China's SRBMs, MRBMs, IRBMs and ICBMs serve as the primary delivery system for China's nuclear arsenal developed solely for deterrence purposes against powerful states like the USA and Russia and to caution other nuclear powers such as India against contemplating the threat or use of weapons of mass destruction (WMD) against China.

China's ballistic missiles range from the 150-km SRBM to the 12000+ km ICBM. These include DF1 (SS2 - SRBM)-range 550 km with 500 kg payload, DF2 (CSS1-MRBM)-range 1250 km with a payload of 15-20 kiloton (KT) nuclear warhead, DF 3(CSS 2- IRBM)-range 2500-2800 km with 2000 kg (or thermonuclear) payload, DF4 (CSS 3-two stage ballistic missile)-range 5500-7000 km with 3 mega ton (MT) nuclear payload, DF 5 (CSS 4 - ICBM) range-12000 km with 3 MT nuclear warhead, DF 11 (CSS 7- road mobile SRBM)-range 300 km (advanced version-825 km) with 800 kg pay load, DF 12 (CSS X-15-SRBM)-range 400-420 km (officially announced due to MTCR restrictions-100-280 km with 400 kg payload, DF 15 (CSS 6-SRBM) range 600 km with 500 kg payload, DF 16 (CSS 11-MRBM)-range 800-1000 km with 1000-1200 kg warhead, DF 21 (CSS 5- MRBM also developed as SLBM)-range 2500 km with 500 KT warhead, DF 25 (two stage solid fuelled IRBM)-range 3200 km with up to three nuclear/conventional warheads, DF 26 (solid fuelled IRBM) range 3500 km, warhead not known (NK), DF 31(CSS 10-road mobile solid fuelled ICBM)-range 11000+km with 1000 KT warhead (or up to three 15-150 KT MIRVs), DF41 (CSS X 10-next generation ICBM)-range 12000-14000 km with single/three/six/ten MIRV warheads.[24]

Besides ballistic missiles, China continues to produce and purchase an array of sophisticated and accurate cruise missiles including the DH-10 and YJ-62 land-attack cruise missiles and the Russian SS-N-22/SUNBURN supersonic air-launched cruise missile. The SS-N-27B/SIZZLER supersonic air-launched cruise missile is placed among China's KILO-class diesel electric submarines purchased from Russia. The YJ-62C, a new variant of the YJ-62, has a range of 150 nautical miles; China has reportedly deployed 120 of these to naval bases near Taiwan.

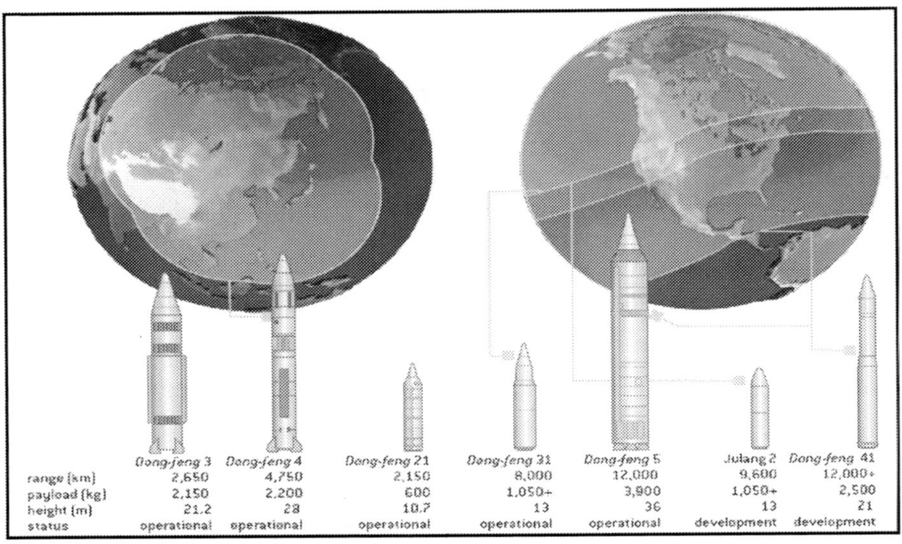

	Dong-feng 3	Dong-feng 4	Dong-feng 21	Dong-feng 31	Dong-feng 5	Julang 2	Dong-feng 41
range (km)	2,650	4,750	2,150	8,000	12,000	9,600	12,000+
payload (kg)	2,150	2,200	600	1,050+	3,900	1,050+	2,500
height (m)	21.2	28	10.7	13	36	13	21
status	operational	operational	operational	operational	operational	development	development

Chinese Ballistic Missiles

The Futuristic Threat from China—Hypersonic Missiles

Hypersonic missiles are those that are capable of flying faster than 5000 km/hr without losing manoeuvrability and can fly lower than the traditional ballistic missiles. Due to these two capabilities, they are very difficult to detect and destroy. This is so because, firstly, by flying lower than conventional ballistic missiles they tend to avoid radar detection and, secondly, owing to their tremendous speed, these weapons provide extremely little time to ballistic missile defences to react and launch the counter in time. Boost and post-boost phase intercepts in the case of hypersonic missiles are remote possibilities. Such missiles are expected to be capable of destroying targets anywhere on the planet within an hour of getting data and permission to launch.

Open sources report that China has been testing the DF 17, a ballistic missile combined with a hypersonic glide vehicle (HGV). This weapon is reported to have been tested on at least at two occasions. In one of the tests, the missile's payload flew to a range of approximately 1400 km before making a successful impact at a site in Xinjiang Province 'within metres' of the intended target.

Since the HGVs fly low, they do not enter space. They come shooting down on targets within the Earth's atmosphere at hypersonic speeds posing a great challenge to early warning satellites and missile defence systems.

According to US intelligence estimates, the DF 17 is expected to reach initial operating capability by around 2020 with the weapon capable of delivering both the conventional as well as nuclear warheads.

Such a weapon poses a great challenge to our ballistic missile defence system.

DF 17

HGV

China tests DF 17 missile with a Hypersonic Glide Vehicle (HGV)

Chinese Build-up in Western Theatre Command

On February 16, the erstwhile seven military regions (MRs) of China have been reorganised into theatre commands (Eastern, Southern, Western, Northern and Central). The Western Theatre Command (WTC) faces India. This command has been created by merging the erstwhile Chengdu and Lanzhou MRs. In WTC what is of special significance is the build-up in the Tibet Autonomous Region (TAR) that directly adjoins India. The Tibet Military Command (TMC) under WTC is responsible for operations in India.[25] As regards the PLAAF, there are three Fighter Divisions (6, 33 and 37) one Bomber Division (36) and one Transport Division (4) in WTC.

Apart from a nuclear missile base in Quinghai province which clearly targets India, China has built five fully operational air bases (Gongar, Pangta, Linchi, Hoping and Gar Gunsa) and an exclusive rail network and over 58000 km of roads in TAR.[26] In 2015, another full-fledged air base at Kashgar located 600 km north of Srinagar was operationalised. An open source reported in April 17 the construction of an airport at Nyingtri close to the borders of India and Myanmar. This civilian airfield is reported to be capable of dual use. The construction of new airfields and up-gradation of advance landing grounds (ALG) is a continuous process in TAR.[27]

The limitations of high-altitude operations with combat aircraft of yesteryears stands diluted to a large extent with the induction of modern aircraft of the likes of SU 27 and J10. To add to this are other 'enablers' like air-to-air-refuellers, AEW aircraft and a strong GBAD/BMD cover centred around S-300/400 missile systems and a hierarchy of air defence missiles.

While a weapon-by-weapon analysis of the Chinese air threat across the entire spectrum is beyond the scope of this chapter, what needs to be done by the defenders in countering the said threat is briefly covered below:

- Capability to carry out effective surveillance of the TAR region with sensor efforts integrated across ground/shore/space mediums resulting in a comprehensive air situation picture.
- A seamless air defence battle management system anchored on auto/real-time data transmission on satellite media with due redundancy.

- Wherewithal and capability to carry out AW counter air operations (CAO), interdiction and counter surface force operations (CSFO) across the border, both in pro-active/pre-emptive as well as retaliatory modes.

Building a degree of survivability in air defence command, control and communication centres both electronically as well as through equipment redundancy.

Building an integrated family of ground, air and shore-based air defence weapon systems to ensure continuous and successive punishment to the air threat right through its ingress into own territory.

For ground-based air defence, the above capability must result in the following:

- Modernised and technologically-enabled gun-missile deployments to cover the entire range-height spectrum from long range to terminal weapons along with their sensors and associated support systems.
- Capability to take on contemporary, as well as, futuristic threats from the PLAAF including stand-off threat and the threat of a slew of smart/intelligent weapons and PGMs.
- Futuristic kill capability in the form of directed energy (laser, CPB and HPM) weapons and fielding a counter for stealth aircraft, ARMs and cruise missiles.
- Building BMD capabilities at the national level to take on the threat of the adversary's SSMs in the manner explained in this chapter.
- In future, the BMD capability must graduate to counter the emerging threat of hypersonic missile systems.

NOTES

1 https://www.skygod.com>quotes>airpower. Accessed on 22 Aug 17.

2 https://www.paf.gov.pk.> Pakistan Air Force - Official Website. Accessed on 30 Aug 17.

3 https://wwwGlobalsecurity.org.> Pakistan Air Force Squadrons. Accessed on 31 Aug 17.

4 https://www. quwa.org>2017/05/21> Pakistan Air Force Squadrons. Accessed on 19 Sep 17.

5 ibid.

6 https;//www.en.wikipedia.org.wiki> integrated guided missile development programme. Accessed on 19 Apr 2018.

7 https;//www.indianexpress.com>pakistan-successfully-test-fires-nasr-shot-range-ballistic-missile-nasr. Accessed on 25 Mar 2018.

8 https://www.nationalinterest.com>Pasistan-could-have-over-100-nuclear weapons. Accessed on 29 Apr 2018.

9 "Nuclear Weapons: who has What at a Glance, Arms Control association. Accessed on 01 May 2018.

10 https://www.orfonline.org>pakistan sea based nuclear deterrent and its asymmetric escalation strategy. Accessed on 01 May 2018.

11 Ankit Panda and Vipin Narang, "Pakistan tests new submarine launched cruise missile. What Now?, *The Diplomat,* 10 January 2017.

12 https://www.quora.com>what are pakistan's nuclear delivery capabilities? Accessed on 01 May 2018.

13 https://www.en.wikipedia.org>national command authority. Accessed on 02 May 2018.

14 https://www.defence.pk>pdf>threads>pakistan's nuclear command and control. Accessed on 02 May 2018.

15 https://www.janes.com> pakistan test fires enhanced version of babur cruise missile. Accessed on 19 Apr 2018.

16 https://www.delhidefencereview.com>does-pakistan's-ababeel-really-have-mikrv-capability? Accessed on 25 Mar 2018.

17 https://www. en.m. wikipedia.org> Peoples Liberation Army Air force. Accessed on 10 Oct 17.

18 http://en.m.wikipedia.org Chengdu J 20.

19 http://www.popsci.com stealthier stealth? Seventh upgraded chinese stealth fighter. Accessed on 10 Oct 17.

20 IHS Janes' International Defence Review, vol-45, Feb 2012, p. 45.

21 https://www.strategy page.com/htmw/ht air w/20070516.aspx

22 Roger Cliff, John Fei, Jeff Hagen. "Shaking the Heaven and Splitting the Earth", Chinese Air Force.

23 Indian Defence Review Vol 30(4) (Oct-Dec 15).

24 https:// www.en.wikipedia.org>wiki>dongfeng missiles.

25 https:// www. Western Theatre Command at https://www.global security.com. Accessed on 13 Oct 17.

26 https://www.times of India.indiatimes.com/2011-03-08/india/28668105

27 https://wwwclaws.in/images/publication.pdf/1317312941MP%2032%20inside.pdf.> China's Infrastructure development in Tibet evaluating trend lines. Accessed on 14 Oct 17.

CHAPTER 2

Some Basic Concepts of Employement

Lt Gen Naresh Chand, (Retd), PVSM

In order to arrive at the present day concepts of employment of air defence resources this chapter tries to trace the evolution of such concepts as these came to be adopted as a result of the 'felt need' of countering the air threat as it presented itself from time to time.

The Evolution Process

The development of warfare over the ages has had two constant factors, that is the evolvement of threat and countering the threat. There are numerous examples of this interplay like 'tank versus anti-tank,' 'submarine versus anti-submarine,' 'shields versus arrows/spears,' 'forts versus attacking cavalry/infantry' and so on. The same is applicable to air warfare which along its journey to becoming a key pillar of military operations, leveraged development of counter measures for it. In this chapter it will be attempted to narrate the play and counter-play of air power and basic concepts of ground based air defence (GBAD) starting with the advent of air power through World War I and II, up to the present jet age. The basic concepts emerge from three components of GBAD, i.e., weapon systems, employment and deployment.

Evolvement of Air Power and Air Defence (AD) Operations

The earlier definition of air power was 'that it conducted operations against other aircraft, army and naval targets.' This definition changed as the development of air warfare unfurled. A more appropriate definition could be the 'capability to project power from the air and space to influence the course of military operations in own favour.' GBAD will also depend upon the characteristics of air warfare.

Before World War I

The American Civil War was fought in the USA from 1861 to 1865 during which the Union Army Balloon Corps was raised and equipped with aerostats to carry out aerial reconnaissance. These aerostats were engaged by ground fire by the Confederate States Army. Similar incidents were reported in Europe even earlier of engaging aerostats with ground fire. Technically, this could be considered the beginning of GBAD.

Army Balloon Corps- American Civil War

The Turko-Italian War of 1911-12, saw the first employment of powered aircraft in war. It took place in the Ottoman province of Tripolitania, which can be generally identified with modern-day Libya. Italy mobilised its Italian Aviation Battalion under the command of Captain Carlo Piazza, a well-known aerial racing pilot. He made history by carrying out the first reconnaissance flight near Benghazi in a Blériot XI on October 23, 1911. Second Lieutenant Giulio Gavotti carried out the first aerial bombardment mission by dropping four bombs on two Turkish-held bases on November 01. In March 1912, another historical first was achieved when Captain

Piazza carried out the first photo-reconnaissance flight in history. The play and counter-play of attack and defence began when the Turks carried out the first ever anti-aircraft operation in history. The first aircraft to crash in a war was flown by Lieutenant Piero Manzini who was shot down on August 25, 1912, and the first aircraft to be captured was that of Captain Moizo on September 10, 1912. Air Defence was known as Anti-Aircraft (AA) in those days equipped with light AA and heavy AA guns. At some stage it changed to AD but it is difficult to give it a date as every country had its own nomenclature. UK probably used the term AD for the first time when it created Air Defence of Great Britain as a Royal Air Force command in 1925.

Turko-Italian War- First use of Powered aircraft

World War I—1914-18 (WW1)

World War I saw the further use of powered aircraft for carrying out aerial reconnaissance which was countered by other aircraft thus making it the beginning of air-to-air combat. As the machine guns carried by the aircraft had a limited range of about 180m (200 yards), the attacking aircraft had to come very close to the adversary to carry out the dog fight. This phase was then followed by the development of tactics for air-to-air combat by all countries possessing aircraft. One example was to 'jump' or 'bounce' a victim from an unexpected direction in order to achieve surprise. These tactics brought more kills than the spectacular aerobatics of a dog fight. It was possible as the pilot had only his eyes as sensors and could look only towards his front. Another tactic was to attack out of the sun thus blinding the adversary, a technique that was used later to attack GBAD weapon systems (GBADWS) which is why gun crews are even today authorised goggles to counter the glare. In those days, flying speeds averaged 160km/h and communication was by hand signalling, wobbling the wings and firing coloured flares etc. But so far there was negligible role for GBADWS which was confined to sporadic fire with existing machine and artillery guns.

It was but natural that the aircraft currently providing aerial reconnaissance moved to the next step to provide support to ground troops by mounting machine guns for strafing and dropping bombs on the enemy. The ground support role was carried out to support advance, cover retreat or just harass the enemy. During the last stages of World War I, ground-attack aircraft had forced the adversary to carry out all large-scale troop movements at night or during bad weather.

World War I - The beginnings of Dog fight

By the end of the war, strong support for the establishment of an independent air force began emerging. The role envisaged for it was to attack deep into enemy territory to destroy/damage the enemy's war-making capability, destroy logistic infrastructure and, possibly, the seat of the government. This role, although never carried out effectively during World War I, was inspired by the German air attacks on London, first by zeppelin airships and then followed by Gotha bombers. These bombers flew at night and often at 20,000 feet to avoid the ceiling of normal fighters. At times, the crew had to carry oxygen bottles to aid breathing. Thus the military role for aircraft which emerged from the World War I was reconnaissance, tactical ground support, strategic bombing and to achieve air superiority. This role for the air force has stood the test of time and has literally remained unchanged.

Basic Concepts for GBAD

As early as 1794, there are records of balloons being engaged by 7-pounder howitzers. Later, the Germans employed 75/80 mm guns in the AA role, the British modified the 13pounder gun for AA role and the French used 75mm field guns for a similar role. This new threat was accepted as the 'threat from the third dimension' and got included in military terminology. The aerial target was complicated to engage as it flew at high speed and could manoeuvre in three dimensions simultaneously. Thus, engaging it with weapons which were designed for attacking static ground targets was completely ineffective. This led to the development of dedicated guns for the AA role. There was an effort by the Krupp Armament Factory in 1870 to build a dedicated AA weapon to take on balloons. The weapon was a 25mm rifle mounted on a cart for mobility. It was nicknamed as 'balloon gun'. There was some effort during World War I to develop AA weapons in Europe and in a way that was the beginning of AA artillery. However, no major development took place during the war, but after the War UK raised training establishments for AA.

Concept

The basic concept which emerged out of the experience of World War I was that there was a need to develop dedicated AA weapons to counter the emerging air threat which may at some stage manifest itself through a separate air force.

World War II—1939-45 (WW2)

The all-metal monoplane was a game changer in terms of performance and firepower over the aircraft of World War I and this change was evident in the fighter tactics.

Air Superiority

Airspeeds of more than 640km/h could be achieved and aircraft could operate up to about 9100m (30,000 feet). The wing-mounted machine guns and aerial cannons had an effective range of about 540m (600 yards); thus there was no need to fly very close to enemy aircraft in a dog fight. Another key development was that pilots could now communicate with each other and the ground control via the radio telephone. The Germans, during the Spanish Civil War of 1936–39 evolved new tactics suited to the greater speeds of their aircraft. This formation was called 'finger-four' In this formation, a group of four aircrafts flew together in a manner which when viewed from the above resembled the tips of the four fingers of a human right hand. There was a "lead aircraft" right in front, followed by two other called "second element" slightly behind while the fourth aircraft was slightly behind the second element. This formation was later adopted by all the major air forces in World War II. Other dog fight tactics were developed which only affected air power and thus are out of the purview of the concepts of GBADWS.

Ground Attack

The German Air Force was primarily organised to support ground forces during the Spanish Civil War and during the earlier years of World War II with Ju-87 Stuka dive bombers as its principal ground-attack craft. The typical tactics followed by Stuka dive bombers was that several aircraft would circle above the target and then, turn by turn, every aircraft would peel away from the formation and dive almost vertically for releasing its bombs, then pull up to join the formation and then repeat the manoeuvre if possible. This tactic was most effectively employed in naval warfare in the Pacific by carrier-based US dive-bombers like Dauntless and Helldiver and the Japanese Type 99 'Val'. The Stuka dive-bombers had a limitation as their top speed during level flight was less than 480km/hr (300 miles per hour) and were thus easy targets for Britain's Spitfires in areas where air superiority was not achieved by the Germans. In such an air environment, Spitfire pilots would have a field day by achieving Stuka kills at will and called this one-sided dogfight as 'Stuka parties.'

Stuka Dive Bomber in Action

Ground attacks were most devastating when conducted by fighter-bombers which often were converted from air-superiority fighters, examples of which are Spitfires and Mosquitos of Britain and P-51 Mustangs, and P-38 Lightnings of the USA. They would fly under the radar horizon to avoid detection, and then carry out bombing of airfields and strafing of infantry columns. The P-51s that performed escort duty for bombers when they were on bombing missions into Germany often attacked ground targets quite freely while returning to Britain while flying at treetop level. In the North African campaign, close air support was fine-tuned by concentrating air power under a joint central authority consisting of senior ground and air commanders in the theatre of operations. This resulted in concentration of fire power launched at decisive targets during the operations.

Concept

The tactic of Stuka bombers diving almost vertically before releasing its bombs was followed so as to drop the bomb as close to the target as possible in the absence of modern electro-optical aids. This was due to the fact that a bomber flying level would have to calculate the point of launch before the target to cater for the speed of the bomber as the bomb inside the bomber will also travel for some distance before it starts to free fall. This implies that the deployment pattern of the GBADWS should be such that it should be able to engage vertically diving aircraft, as well as, aircraft that attacked from other angles including level flights.

Dedicated GBADWS were further developed like the German WW2 flak guns which Germany employed along with all types of AA weaponry throughout World War II and even captured enemy systems to defend the skies over its new-found territories. Along with the Swiss 20mm Oerlikon, the Swedish 40mmL60 Bofors became one of the most successful light anti-aircraft weapons in military history.

Introduced in 1932, the Swedish Bofors 40mm L-60 AA gun was the most effective gun for both land and naval roles during World War II. It was in service in India up to the 1990s till it was replaced by 40mm L-70 which is also from the same company. In 1938, Britain produced a 3.7 inch AA gun which weighed 9320 kg, fired a 28-pound shell to an effective ceiling of 2300 feet at a rate of 10 rounds per minute. This gun was used extensively in World War II and also saw service in India where it was used as an interim medium artillery gun till 130mm medium guns were imported from Russia. This gun was also connected to a Predictor (forerunner of current computers) which worked out the miss distance of a moving aerial target.

The other aspect that emerged was to develop low-looking radars to detect targets flying at tree top level. It was also essential to augment radars with observers.

Strategic Bombing

Massive bombing of military targets and major cities took place during World War II, causing mass destruction of a scale that is unlikely to be repeated in future wars unless a nuclear strike is carried out. The bombers were slow moving and huge and thus exposed themselves as easy targets but they came in large numbers, sometimes as large as 1000. The aim was not to avoid enemy defences but to saturate them with numbers and fire power. Bombing during the day required a fighter escort which provided safety from GBADWS as they were beyond its reach. A key example of air battle and the development of an integrated AD system took place during the Battle of Britain.

Battle of Britain

Battle of Britain started on 10 Jul 1940 when Germany launched massive pre-emptive strikes on Britain in order to destroy its infrastructure and air power. German aircraft attacked coastal targets and British shipping in the English Channel. Germany's main offensive was launched on August 13, in which air attacks took place on targets in depth like airfields and communications centres. But unfortunatly for Germany, its air force over-estimated the damage it had inflicted on Britain and erroneously concluded that the Royal Air Force (RAF) was on its last legs. Basewed on this assumption, concentrated all its effort on London. This pause gave time to the RAF to recover and on September15 Britain's Fighter Command repelled a massive attack by Germany and inflicted heavy casualties to their Air Force. It gradually dawned on the Germans that their attacks are not sustainable due to their heavy losses and thus had failed to achieve air superiority, so essential for launching Operation 'Sea-lion'(invasion of Britain) resulting in its postponement indefinitely. The battle received its name from Winston Churchill when he said in the House of Commons on June 18, 1940, that, "the Battle of France is over. I expect the Battle of Britain is about to begin."

Battle of Britain

British AD Organisation

The RAF evolved an elaborate AD system to defend Britain from the German Air Force. This system gave them a key edge over the Germans in the Battle of Britain. It was called the Dowding System named after the Fighter Command's Commander-in-Chief, Air Chief Marshal Sir Hugh

Dowding. The system unified technology, ground defences and fighter aircraft into an integrated system of AD. Britain was divided into four geographical areas called 'Groups' which were further divided into sectors. The main fighter airfield in each sector had an operations room to direct all elements involved in the AD system. Radar had already been developed and employed during World War II. Early warning was given by the radars and the Observer Corps (1,000 observation posts were established) to scramble the fighters; alert AD guns, searchlights (for night) and barrage balloons. The Dowding System was connected by elaborate communications and could process early warning data and ensured that the limited AD resources were not wasted. Apart from very good fighter pilots, excellent military leadership in Air Marshal Hugh Dowding and an elaborate AD system gave a critical advantage to the RAF over the German Air Force. This AD System can be considered a first for AD operations.

Concepts Emerging from the Battle of Britain

- For providing effective AD, an integrated system was required which included all the elements required for an AD battle.
- The elements of the system included fighters, sensors and GBADWS.
- Fighters, sensors and GBADWS should also be able to operate at night and in bad weather conditions.
- The weapon systems should be able to provide a gap-free AD by day, night and in adverse weather conditions.
- Warning of approaching enemy aircraft was a key element of the Dowding System.
- A command and control centre was required to integrate the key elements to ensure adequate warning of air attacks and judicious allocation of resources to affect economy of effort. All the elements should be elaborately connected with radio communications. The command and control centre should be able to operate round-the-clock.

Post Battle of Britain Period

The RAF used two radar-beam systems called Gee and Oboe to guide its Lancaster and Halifax bombers to cities in Europe since late 1943. In addition, the bombers carried a radar mapping device, code-named H2S that helped the navigators in finding the target area. The RAF aircraft dispensed very large quantities off tinfoil strips called 'window,' to jam German radars. This can be considered the beginning of Electronic Warfare (EW).

Another important aspect was bombing at night as it made the enemy fighters ineffective. This later on changed with radars coming on board the fighter aircraft. It however remained very difficult to locate and engage a target except on clear moonlit nights with visual navigation aids. The advent of electronic aids made night air operations easier. The Germans employed a unique system called Knickebein in which bombers followed a radio beam being broadcast from ground stations in Europe until the beam was intersected by another beam at a point over the target. Lead bombers then dropped incendiary bombs to start fires that guided other bombers carrying high explosives, as well as, more incendiary bombs.

Concept

This further reinforced the requirement of GBADWS being able to operate during the night, as well as, in adverse weather conditions.

The requirement of electronic means of surveillance emerged much before World War II. Scientists in many countries like Britain, France, Germany, Italy, Japan, the Netherlands, the Soviet Union and the USA, independently and under great secrecy, developed technologies that led to the modern version of radar. Britain developed an elaborate radar system for its defence which was called the Chain Home (CH) system.

By 1936, the first five CH systems were operational and by 1940 it covered the whole of Britain including Northern Ireland. CH was a simple system which, instead of transmitting and receiving from an aimed antenna, transmitted signals literally floodlighting the area in front of it and employed Watt's own radio direction finders to determine the direction of the returned echoes. This simple system was easily manufactured and deployed in a short time instead of waiting for more sophisticated systems to be developed.

The Battle of River Plate was the first naval battle in World War II and the first of the battles in the Atlantic in South American waters. During the battle, the German ship, Admiral Graf Spee, was scuttled in the River Plate estuary on December17 by its captain to avoid unnecessary loss of life. A radar expert was sent to Montevideo by Britain shortly after the scuttling and reported a rotating aerial transmitting on either 57 or 114 cm and could probably be used for gun-laying. This indicated that the Germans were also quite ahead in the development of radars. Further development continued for better resolution, more mobility and more features for radar, including complementary navigation systems like Gee and Oboe used by the RAF's Pathfinder. From such humble beginnings, radar has developed into a critical component of GBADWS.

The RAF gave birth to EW when its aircraft started dispensing very large quantities of tinfoil strips called 'window,' to jam German radars. Thus GBADWS had to counter the jamming threat from the air and also try to jam them. This has led to advanced countries like the USA having dedicated EW systems including those mounted on aerial platforms.

The Jet Age

The end of World War II saw the first operational jet fighter; the German Me-262, which was much faster than the best Allied aircraft. This aircraft ushered the dawn of the jet age in which aircraft soon flew at more than twice the speed of sound and comfortably climbed to 50,000 feet. Simultaneously, avionics was developed to provide surveillance, navigation, weapon control and communications. The range of aerial combat was extended with the introduction of guided missiles. The march of development led to greater speeds; improved avionics including head-up display and Active Electronically Scanned Array (AESA) radar; stealth; all-weather capability; aerial attacks coupled with electronic attacks and stealth and mid-air refuelling. All this brought about changes in air tactics and matching GBADWS capability.

Suppression of Enemy Air Defence (SEAD)

During the 1960s, it was evident that the radar-controlled GBADWS were a serious threat to attacking aircraft. Tactics like flying low and fast, jerking by making quick, irregular changes in direction and speed or diving in a steep spiral did help in avoiding GBADWS but then the mission had to be aborted. Thus it became clear that the GBADWS were to be deterred/ destroyed by superior speed, outgunned as well as electronically jammed so that their surveillance and guidance sensors become inoperative. This was termed Electronic Counter Measures (ECM). The follow-up aerial missions could then carry out successful attacks under the shadow of SEAD missions.

There are examples of this in the Vietnam War when the follow-up missions came under a massive cloud of chaff to blind the Russian origin guns and missiles. Radar homing missiles were also developed by the USA during this period which were launched by F105s to destroy Vietnamese radars. During the Arab-Israeli War of October 1973, Israel lost 40 aircraft to Egyptian and Syrian AD in the first two days. Israel developed new tactics and in June 1982 destroyed Syrian SAM sites in al-Biqāᶜ Valley, Lebanon.

In that, massive pre-emptive electronic strikes were carried out by dedicated EW aircraft to confuse the Russian-origin missile sites. This was followed by launching of small UAVs by Israel which were attacked by Syrian defences and thus exposed them to be destroyed by radar-homing missiles and cluster bombs of the Israel Air Force. The requirement of SEAD has been reinforced during all subsequent air operations like the air offensive during Operation Desert Storm.

Other Aerial Platforms

Rotary wing aircraft have become a potent aerial platform when armed and have proved very effective in anti-personnel and anti-tank roles. Further, the development of unmanned aerial systems (UAS), and missiles (cruise, tactical nuclear, ballistic) have added a new dimension to the air threat. It depends on the country as to how it perceives its security environment to contain the air threat. For example, countries like the USA consider intercontinental ballistic missiles as a threat and are developing systems to counter them. So also are India and China.

Air Threat

The evolution of air power has been narrated from the period when it was first employed to present times. Concepts for GBADWS have also been deduced wherever applicable. An effort will now be made to consolidate the current air threat at the conceptual level and infer basic concepts from it for the selection, employment and deployment of GBADWS. It may be emphasised that every country has to assess its own air threat depending upon its regional security environment.

Basic Concepts

Air Threat: Air threat can be considered in three parts, i.e., qualitative threat, quantitative threat and attack profiles. The platforms which can be included are fixed-wing aircraft, rotary-wing aircraft, UAS, armed UAS, cruise missiles, tactical ballistic missiles and surface-to-surface long-range missiles.

Qualitative Threat: This gives the specifications of the aerial platforms like speed of attack; accuracy of attack; quality of guidance systems; degree of stealth; on-board protection systems; electronic jamming capability; low-level flying ability; radar avoiding capability; on-board munitions for fixed wing aircraft, rotary wing aircraft and armed UAS including beyond visual range(BVR) engagement capability, joint direct attack munitions (JDAM) and joint stand-off weapon (JSOW) which make use of the global positioning system (GPS) of the USA for guidance. Each aerial platform has to be analysed separately and suitable GBADWS designed/selected to counter it. Fixed-wing aircraft, rotary wing aircraft and UAS will also carry out intelligence, surveillance and reconnaisance (ISR) missions which may require separate GBADWS to counter them. It may happen that some GBADWS may be able to take on multiple aerial platforms.

Quantitative Threat: This gives the number of aerial platforms that are coming for the mission. The simplest example is of a fighter mission which may have four or six aircraft or more in each mission. For missile attacks, it could be two missiles or more. This will give out the quantum of GBADWS required to simultaneously deter or destroy aerial platforms attacking in a single raid. (the above figures are only illustrative)

Attack Profiles: This will give the tactics employed by the aerial platform like height of approach, attack and getaway which is applicable to fixed-wing aircraft, rotary-wing aircraft and UAS. The profiles are described as "hi" or "lo" related to the three stages of a typical move of an aircraft coming in for attack. These stages are "approach", "attack" and "getaway". For instance there could be a profile, hi-lo-hi or lo-lo-lo etc. The attack profile will affect the deployment pattern of GBADWS, e.g., a lo-lo-lo profile would mean GBADWS including their sensors should be able to detect and engage low-flying aerial platforms. Cruise missiles will fly hugging the terrain and other missiles will have their own attack profiles. All GBADWS should be so deployed so that they can engage the mother platform before it can launch its weapons. In the case of air-to-ground missiles, GBADWS should be so deployed to destroy them before they can reach their target.

Attack Envelope: The above threat parameters will give out a threat envelope around the defended targets, both horizontally and vertically. This leads to the selection of a suitable GBADWS to counter each area of the threat envelope.

All round Defence and Defence in Depth: This is the basic principle of all defensive layouts and also applicable to GBADWS. It implies the ability to counter an attack from any direction and successive GBADWS should be able to engage it so that if one system misses it, others can take it on.

Scaling of GBADWS: The scaling of GBADWS is a difficult task because GBADWS have high cost and an aerial platform is very flexible in employment. Theoretically, a couple of fighters can attack any target in the area and to counter them GBADWS may have to be provided to defend numerous targets. The other aspect is the quantum of weapons required to destroy an aerial platform. The Russians had earlier suggested six guns (radar control or manual) or two missiles for one target. This is an overkill and with modern sensors, radar control guns with better ammunition and advance missile systems, a certain amount of economy can be effected depending upon the air threat and the priority accorded to the asset being provided AD.

The Current Concepts of Employment

Having seen the basic concepts of employment of air defence resources and their evolution over time, this portion analyses the following two aspects which will lead the reader to the current concepts (read necessities) of employment of air defence resources. Much of this analysis draws from the content on the evolution process of various concepts as covered in the preceding paragraphs.

1. Why guns alone are not adequate?
2. Why an integrated family of air defence resources is a must?

Why Guns Alone are not Adequate?

As stated earlier, air defences across the Services exist for the sole reason to protect own selected vulnerabilities from the air threat of the adversary. The qualitative and quantitative developments in the severity and the lethality of the said air threat, over time, is tied to the development (read necessity) of GBADWS in a cause-effect relationship. In this reciprocal relationship, one drives the other and vice-versa in an eternal duel much like the dictum—"to every sword a shield".

It therefore follows that the answer as to why SAMs, why such a variety of SAMs and why not guns alone for the GBAD requirement is therefore to be found in the metamorphosis in the air threat that has truly revamped itself exponentially with the march of technology. As explained in detail in the evolution of concepts over time, this process of metamorphosis is briefly recaptured below:

- There was a time around the late Sixties when the air threat was mainly prosecuted by the binary duo of aircraft and helicopters, both fair-weather machines of first/second generation with such munitions as front guns and unguided bombs and rockets.

- On the other side, the vulnerable areas/points (VAs/VPs) to be protected by GBADWS were limited and essentially required fair-weather protection from the above-said threat vehicleslargely executing their end game in the visual domain.
- The resultant trend was to deploy multiple rings of terminal AD guns (which were considered a good fit) around the vulnerabilities to be protected in the 'point defence' mode or go for 'carpet layouts' (increasing densities of guns) where the VA/VP was that critical.
- As time rolled on two dimensions of the air threat became increasingly visible. First was the multiplicity in the air threat vehicles and the second was the exponential rise in their range and reach envelope. In that, the erstwhile binary pair of aircraft and helicopters were joined by unmanned aerial vehicles/unmanned combat aerial vehicles (UAVs/UCAVs) besides ASMs/cruise missiles/ARMs and more. Also, as time passed, the technology made it possible to take quantum jumps in all-weather capability, longer ranges, deep strikes with precision and standoff capabilities of the attackers. In short, the sting of severity and lethality in the air threat vehicles got a paradigm upward shift over time.
- The above trend of revamping of the air threat continued exponentially over the years and mere guns alone with just a few km of range and altitude capability proved to be totally deficient in taking on multiple all-weather threat vehicles with precision kill munitions.
- For giving successive punishment to the incoming threat, the need for such GBADWS was felt which could be deployed along the entire range and altitude brackets and which could inflict debilitating punishment to the incoming hostile targets from very long ranges to close domain. This requirement could only be met by longer range and reach weapons like the SAMs.
- This addresses the first query - why guns alone are not adequate.

Why an Integrated Family of Air Defence Resources is a must?

The answer to this poser is to be found in the challenges for air defence warriors that grew exponentially over time. A snapshot...

- The sheer numbers of VAs/VPs saw a tremendous exponential rise as more and more assets came under the 'attack-able reach' of constantly re-vamping air threat vehicles. It soon became impossible to protect each one of them in the 'point defence' mode.
- 'Point defence' as a concept not only proved impossible because of the sheer number growth in VAs/VPs, but also, because it proved totally ineffective in the face of all-weather, multiple platform, multiple-layer, multiple-arsenal air threat delivered with needle-like precision at far stand-off ranges in the BVR domain.
- The defenders realised that instead of point defence of single VPs the need was to create an area air defence capability on a theatre grid. Such an arrangement must be capable to detect the threat at long ranges and be capable of inflicting punishment to it through multiple GBAD weapons in successive range and altitude brackets from long to medium to short and terminal ranges.
- From the above felt need was born the concept of layered-and-tiered defence. The idea was to create a theatre-wide defence through long range (LR), medium range (MR) and short range (SR) SAMs referred to as LRSAMs, MRSAMs and SRSAMs, respectively, creating overlapping arenas of range and altitude kill zones with seamless switching from one weapon system to other.
- It also got established unambiguously that keeping the current and the futuristic air threat in mind, whatever may be the range and depth of SAMs that may be deployed at higher ranges and altitudes, the end game at the terminal end was still critical and needed to be executed with quantum adequacy of weapons and modernity of arsenal.
- Also, since all VAs/VPs (due to sheer numbers) could not be given point defence by a finite number of terminal GBADWS as were available, the ones given AD, must be "adequately protected". For this to happen, single weapon (implying guns alone) defence at terminal end was found to be highly inadequate. The minimum required was a combination of guns and very short range air defence

systems or VSHORADS. This was referred to as the 'gun missile mix,'

- In addition to the above, a special type of VPs also needed to be protected. These were the mobile mechanised (armoured) assets operating in the fluid battle in the tactical battle area (TBA). These were specific to field forces. For them, two other GBADWS types came in over time, namely, AD gun missile self propelled or ADGM (SP) and quick reaction SAMs or QRSAMs.
- As technology advanced, the base concept of layered and tiered defence, comprising guns and SAMs duly supported by state-of-the-art sensors and associated battle management command and control (BMC2) systems were networked across dissimilar systems evolved as a cumulative punch called the integrated air defence system, or IADS for short.

  In the IADS system stated above, SAMs of varying ranges and altitude capabilities are essentially required, each for its specific role in countering the contemporary air threat of today—the 21 century air threat. What is the 'feel' of such a threat today?

Well, the same is being prosecuted by the fifth generation fighter aircraft or FGFAs (sixth generation not far) in growing concert with the dull, dirty and dangerous unmanned machines (complete with manned and unmanned teaming or MUM-T. The threat features all-weather, all-aspect stealth, nano-driven structurally lighter and stronger airframes with minimal radar cross-section (RCS), common data links (CDL) providing seamless connectivity across dissimilar platforms, high survivability in hostile EW environment and armed to the teeth with precision arsenal capable of taking out the proverbial 'needle in the haystack' from increasing stand-off ranges.

This addresses the query why IADS?. To sum up, following concepts of employment emerge:

- To effectively counter the contemporary, as well as, the futuristic air threat an integrated family of GBADWS is required consisting of guns and a hierarchy of SAMs starting from very short range to short range to medium and long range.

- A combination of sensors (surveillance and fire control systems), shooters (guns and missiles) and a fully packaged battle management system (capable of seamlessly integrating similar and dissimilar systems) is required to form an IADS which also has, besides the GBAD, the air defence means in the air and at sea.

CHAPTER 3

Battle Management of Air Defence Systems

Lt Gen (Dr) V K Saxena, (Retd), PVSM, AVSM, VSM

A typical air defence battle has some uniquely characteristic features which set it apart. These features actually become the terms of reference which define the confines under which the air defence warrior must aim to defy and negate the air threat at hand. Since the initiative in the air defence battle resides with the attacker, the said terms of reference are also set by him. He does it by setting the battle environment with his capability to prosecute the air threat through multiple air threat vehicles comprising aircraft, attack helicopters (AH), unmanned aerial systems (UAS), cruise missiles anti radiation missiles (ARMs) and more duly armed with a slew of smart and intelligent munitions, operating both in the conventional, as well as, the soft kill domain. This offensive punch is further revamped by the enabling technologies like stealth, nano artificial intelligence and more. All these combined, provide the attacker with a resultant capability of delivering a debilitating lethal punch with speed and precision.

Such a battle environment set by the attacker poses a world of challenge for the defender. This challenge is actually contained in the three words: 'search,' 'identify, and 'counter'. Each one of these words is a vertical in its own right. The bottom-line is, that the entire cycle of the air defence battle to counter the threat as described above, needs to be completed in a matter of just a few fleeting minutes.

Chinese J 20 Super Stealth Fighter

The responsibility to do the above task lies in the domain of the air defence battle management command and control (ADBMC2) system, also referred to in part as the air defence control and reporting system (ADC&RS). The cardinal importance of ADC&RS is contained in the fact that leaving aside the actual endgame of countering the threat by launch of weapon systems everything else leading up to it, lies in its forte. Simply put, if the ADC&RS muscle in an integrated air defence system (IADS) is weak, no air defence weapon system of any sort can perform optimally. Even the best of technology will simply prove to be ineffective if the ADC&RS is not up to the mark. This puts the relevance of this chapter in the right perspective.

ADC&RS draws its pivotal importance from the continuum of activities it controls in the air defence battle chain. A brief overview of such activities is attempted.

The first is the task of initial detection of the air threat vehicles in the air space followed by their identification as being friendly or hostile (referred to as identification friend or foe or IFF). These two by themselves are huge technological challenges. Due to the availability of multiple threat vehicles with the attacker, a large number of them are launched in airspace at the same time. These not only need to be discerned as to their type and likely intentions but also to be identified within a few moments as friendly or hostile. Also, since multiple sensors (radars) of the defender look at the huge array of aerial threat vehicles of the attacker at the same time, there is a challenge of fusing their inputs in real time so as to cut out duplication in reporting. This activity is called multi-sensor fusing or MST for short.

Post detection and identification comes the task of prioritising the incoming threat as to its comparative lethality with reference to the vulnerabilities being protected. For this, against a huge array of variables related to the demonstrated and implied behaviour of each air threat vehicle, its likely lethal payload, intentions and more, as applied to own assets need to be processed in an automated mode through a man-machine interface.

Based on this priority, the system then makes a selection of the air defence arsenal (aircraft, SAMs, terminal weapons of Air Force, Army and Navy) to inflict punishment for effect. This selection is not rigid or one time; it is seamless and dynamic. Seamless, because the selection of weapons for fire cuts across Service The criterion is to maximise the effectiveness at target end by choosing the most suitable weapon system at the instant of application, irrespective of the fact as to which Service it belongs to. Dynamic, because the initial selection is subject to change in real time from weapon-to-weapon and from GO to NO GO in response to the dynamics of the changing behaviour of the incoming threat and safety of our aircraft engaged in combat; the bottomline being to inflict successive seamless punishment on the threat as it draws inwards from long ranges to its intended target. This prioritisation, selection and control of fire is referred to as minute-to-minute control of the air defence battle.

Air Defence battle management—a key battle function

It will be noticed that if there is one thing that is 'fleeting past' in this whole cycle, it is TIME. Invariably, the entire sequence, right from identification of the threat to the initial scramble for interception and going up to the end-game of terminal end engagement may just extend to a few MINUTES only. This defines the criticality of the ADC&RS system and the need for its super high response capability in executing the battle functions enumerated above. This also explains why without an effective ADC&RS, any air defence weapon will become suboptimal.

Besides the above battle sequence, ADC&RS deals with a hundred other things like station keeping, resource management, auto status updates and, above all, airspace control (which is a huge vertical by itself). Also the stealthy targets, which 'never show up' but which deliver their lethal load at long stand-off ranges have also to be dealt with by controlling out-of-sight engagements.

By compulsion therefore, the system to execute the above challenge has to have a tremendous response. It must be fool-proof, redundant and near real-time. Countries around the world are fielding complex ADC&RS featuring the latest technologies in surveillance, data transmission and battle control. In specific terms, a typical ADC&RS must have certain characteristic signatures, some of which are enumerated below:

- A complete array of control centres (called ADC&RS communication nodes) established all across the area of responsibility.
- A main spinal artery of communication highway normally based on redundant media that connects the above ADC&RS nodes with a whole lot of sensors (radars) deployed upfront as to cover maximum range and depth across the borders, as well as, with sensors that are with the air defence weapons from the theatre to field level, cutting across Service boundaries.
- A redundancy to the main communication spinal with a capability of auto-switching of media in a dynamic and flexible manner.
- Add-on systems and organisations to execute associated functions, viz., IFF and Air Space Control.
- Inbuilt system capabilities for optimal resource allocation, task ordering and management of different threat requirements across diverse range and height bands.

All the above make up for a complex system. Challenges like tying up dissimilar air defence systems in one command and control loop, achieving a perfect multi-sensor fusing, or a 100 per cent foolproof IFF, are something the world is still grappling with. What is the Indian scene? To the extent open sources permit, this much is known:

- An integrated air command and control system (IACCS) is in place for controlling and monitoring air operations by the Air Force at strategic operational and tactical levels cutting across Service boundaries.[1]

Integrated Air Command and Control System – IACCS

- The system has a connect with a large array of sensors belonging to both Service and civil domains besides human intelligence resources or mobile observation posts (MOPs).
- The communication and data transfer backbone called the Air Force Net or AFNET is a satellite communication (SATCOM) wide area network (WAN) with internet protocol (IP) based on open architecture and state-of-the-art human-machine interface (HMI). IACCS operations will ride the AFNET (launched on September 14, 2010) integrating all ground-based and airborne sensors, air defence (AD) weapon systems and command and control (C2) nodes. Subsequent integration with other Services networks and

civil radars will provide a recognised air situation picture (RASP) to operators to carry out AD role.[2]

- As stated earlier, the system is capable of generating accurate, updated and complete RASP for local, as well as, centralised air space control, surveillance and threat evaluation on the basis of identification of tracks. It can provide weapon allocation solutions to aircraft for interception and to other ground based air defence (GBAD) means for engagement of targets.
- The Government has cleared a proposal for allocation of nearly Rs. 8000 crore for IACCS. This fund injection will take the IACCS towards completion with a comprehensive capability to get the RASP, integrate different sensors of the three Services and the ones in the civilian domain and establish all the required communication nodes or sector HQs.[3]
- Till date, five communication nodes located at Barnala, Wadsar, Aya Nagar, Jodhpur and Ambala have been established by the Air Force under the IACCS. In Phase 2 of the IACCS, four major nodes and 10 sub-nodes are planned. Three of the four major nodes are one each for the eastern, southern and central sectors, while the fourth one is for the Andaman and Nicobar Islands Project.[4]
- Besides satellite connectivity with ISRO-launched satellites (GSAT 7A), the plan is to continuously leverage cutting edge space based technologies for IACCS in the field of surveillance, reconnaissance and battle management and control as these keep becoming ripe for deployment in the ISRO's march into the future

Alongside IACCS, the Army's system for ADC&RS responsible for the tactical and minute-to-minute control of the air defence battle as prosecuted by GBAD weapons under the jurisdiction of Army Air Defence has also been designed. Similar is the case with the Navy's ADC&RS system.

What IACCS does for the Air Force sensors and shooters, the Army's ADC&RS does for the Army AD, though on a lower scale and scope. Through IACCS connect, the Army ADC&RS is designed to receive multi-sensor fused inputs, near real time data transfer, latest and real time RASP and more, against a geo information system or GIS background.[5, 6]

The basic building blocks of Army's ADC&RS run along similar lines as that for IACCS. The media for battle control and data transmission is similar SATCOM established on a WAN with IP protocols. Keeping the mobile nature of the Army air defence weapons and operations in mind, mobile SATCOM terminals have been designed.

These Army ADC&RS nodes are deployed alongside various ground based air defence weapon systems (GBADWS) in the field and control weapons placed in their tactical and operational control under the overall jurisdiction and control of IACCS.

Since the air defence battle does not recognise nor differentiates between Service boundaries, it is one seamless onslaught of lethal punishment that has to be inflicted upon the aerial attacker, right from the time it is detected hundreds of kilometres away till such time as it is destroyed or the mission abandoned. All this, has to be done in a matter of a few fleeting minutes as stated earlier.

Since the weapons that have to execute this battle function reside in the core competency domain of different services (Air Force, Army, Navy, Coast Guard), the ADC&RS of each Service controlling its air defence weapons must be integrated in one seamless loop of communications and connectivity so that the execution of the air defence battle may pass seamlessly from weapon-to-weapon and system-to-system regardless of the fact of ownership (and hence control) of such weapons by an individual Service.

Such seamless connectivity is not only required for the effective execution of the air defence battle but also to avoid fratricide. Blue-over-Blue continues to remain the unpardonable sin of any air defence warrior. This integration essentially means the following:

- Media commonality not only in generics, but also in versions, generations and levels between the two systems
- Seamless flow of RASP from the highest ADC&R node to the lowest in a selective presentation mode, wherein, a typical ADC&R entity, based on its level and authorisation, is presented only that part of the RASP as relevant to it, in real time.

- Macro-micro in the same node at the same time. This implies, that the relevant theatre information, as well as, the one that is relevant to the effectiveness of the weapon(s) system controlled by a particular C&R node is presented in a PIP (picture-in-picture) mode in the same node with a capability of seamless transition of incoming threat from the macro to micro display as the threat draws towards the asset controlled by the node.
- Real time flow of track-ridden IFF information, authorisation/ prohibition of fire, airspace control instructions post auto-conflict resolution and dynamic upgradation of base details in terms of fire areas, no fly zones and more.
- Auto-switching of media in case of catastrophic damage/force-majeure situation.
- Human override where safety so dictates.

Last, but not the least, there is this all important issue of survivability of the system in an electronically hostile environment. It is to be expected that the adversary will be prosecuting the air threat along with a strong muscle of electronic warfare (EW). In that, there will be jamming attacks on radar systems, control nodes and data flow highways, all contributing towards the overall efforts for suppression of enemy air defence (SEAD) operations. It has therefore to be ensured that the entire ADC&RS system has adequate survivability which may manifest as under:

- Latest electronic counter counter measure (ECCM) features built into the surveillance systems, communication and control nodes, as well as, data flow highways.
- Redundancy in media with a provision for auto switching in case of debilitating EW attacks by the adversary.
- Multiple "work through jam" procedures built into the system as standing operating procedures (SOPs).
- Provisions for manual override for critical connectivity in case of catastrophic failure with built-in provisions for auto indication and auto switching.

NOTES

1 https://www.indrastra.com>analysis-integrated-air-command-and-control-system-a-ncw-milestone. Accessed on 13 May 2018.
2 https://en.wikipedia.org/wiki/air-force-networkAccessed on 20 Dec 17.
3 http://economictimes.indiatimes.com/ government-clears-rs-8000-crore-iafs-integrated-air-command-control-system.Accessed on 20 Dec 17.
4 https://www.indrastra.com>analysis-integrated-air-command-and-control-system-a-ncw-milestone. Accessed on 13 May 2018.
5 www.bel-india.com/?q=ADC-R. Accessed on 23 Dec 17.
6 www.bel-india.com/?q=c4i. Accessed on 23 Dec 17.

CHAPTER 4

Air Space Control: Challenges and Way Ahead

Air Marshal Anil Chopra, (Retd), PVSM, AVSM, VM, VSM

In generic terms, Air Space Control (ASC) refers to a set of instructions that regulate the use of airspace by multifarious users be it military or civil. The need for regulation obviously arises due to the fact that the finite resource of 'airspace' is required to be used by multiple users at the same time. The bottom line requirement is to ensure maximum freedom of action for each user with minimum restrictions, while at the same time keeping the safety of air traffic intact and ensuring that the fratricide is kept at bay.

This Chapter visits the issue of ASC in all its dimensions, starting from the genesis of the growth of air power and the resultant growth of the air threat vehicles that proliferate the finite air space today.

Inter-continental operations became possible in World War II because of maturing of air power which in a great way decided the outcome of the six-year war. More recently, the Arab-Israeli wars of 1967 and 1973, Bekka Valley operations, the Falklands war, the Bosnia conflict and the post 1990 wars in the Middle East, Afghanistan and Libya have been predominantly air wars. Closer home, the Indian Air Force (IAF) played a significant role in the victory in 1971 and Kargil Indo-Pak wars. The coordinated 9/11 aerial attacks against the USA by al-Qaeda have brought a new dimension to the air threat. Today, the fighter bomber has become faster, more agile and stealthy. Transport aircraft have a global reach. Helicopters in tactical battle area (TBA) are using Nap-of-the-Earth (NOE) techniques. The proliferation of inexpensive unmanned aerial systems (UAS) have added a cheap but a potent weapon in the hands of many. UAS proliferation at various levels in the armed forces and with huge civil applications makes a dynamic change to the aviation environment and needs fresh rules to be laid out. The world is engaged in developing counters to ballistic and tactical missile threats. The threat of an aerial attack launched from space today is real. The air threat essentially includes aerial platforms, space-based launch platforms and surface/air launched weapons. Meanwhile there is an exponential increase in civil air traffic.

Exclusiveness to military air space is shrinking. Plans are afoot for space tourism. Air taxis are a reality. UAS has also begun playing a great role in day-to-day product deliveries and also for aerial policing, among many other roles. The Chinese have flooded the markets with hand-held UAS for hobbyists. The density of air traffic is thus increasing at a very high rate. With this intense environmental background, military and civil aviation have to co-exist without severely hampering the efficient operations of either. Air space control thus poses fresh challenges and is constantly looking for fresh solutions.

As stated above, before dealing with the issue of air space control (ASC) per se, a genesis as to how the finite air apace is getting congested by multifarious threat vehicles making ASC a major issue has been attempted. This has been done by sketching in brief the current air threat scenario.

Air Threat to India

At the strategic and tactical levels, China's air power can now achieve a variety of effects. Though their current concentration is on operations on the eastern seaboard in the South China Sea, the same weapons will be used against India. China wants to exploit the advantage of using its tactical/ strategic missile force, which is easier to use for offensive than defensive. The People's Liberation Army Air Force (PLAAF) plans to move the forward edge of the battle into the enemy territory. It will use its air offensive to keep the IAF head-down. Like the Indian Air Force (IAF), the PLAAF has switched to net-centric offensive air defence placing greater reliance on integrated attack. China's ambition is to build its airpower like the USA for an asymmetric advantage. It is aiming to be one of the world's foremost air forces by 2020, made up of at least 1,000 'modern' combat aircraft.

Russia delivers first four SU 35s to China

The Russian Su-35 aircraft, along with its advanced IRBIS-E passive electronically scanned array radar system is likely to enter service by 2018 and will significantly enhance the PLAAF's ability. Of greater concern is the offensive capability in terms of PGMs and the surface-to-surface missiles. With 20 combat squadrons having 450 combat and over 300 other support aircraft, the Pakistan Air Force (PAF) remains significant, but it is heavily dependent on China for all hardware and support. The F-16, JF-17 and FC-20 will finally be the main types. Pakistan has been in talks with China to acquire JF-31 Stealth fighters and with Russia for Sukhoi Su-35 air-superiority multi-role fighters. Being an air defence centric air force, the PAF has little offensive capability and will not be much threat to India. Pakistan has an evolving surface-to-surface missile force and India needs to cater for it.

MILITARY AIR SPACE DIMENSION

Military Aerial Platforms

The fighter-bomber aircraft remains the main instrument of prosecuting air war and conversely also for air defence. In addition to creating air superiority for unhindered operations of surface forces, they have the capability to deliver very lethal and accurate aerial weapons and the fighter fleets consume major portion of defence budgets. Their main characteristics are agility, super cruise, stealth, multi-function AESA radars, network-centric systems, integrated glass cockpits, fibre-optics data-transmission, multi-spectral sensors, fused situational picture, helmet mounted sights and precision guided weapons (PGW). Fighters strive to have 'first-look, first-shoot, first-kill' ability. Other significant airborne platforms that support air operations are the airborne early warning and control systems (AEW&C), electronic warfare platforms and aerial refuellers.

The special operations aircraft can induct Special Forces and cause havoc. Transport aircraft are used for inter-theatre movement. Helicopters have a great role in tactical attack, battlefield logistics, surveillance and casualty evacuation among others. More and more of these roles are gradually being taken over by unmanned or optionally-manned aircraft. Most army units have hand-held UAS. A large number of surface-to-surface missiles (SSM) are in the TBA. Sometimes, long-range missiles may be used. Each of the armed forces has surface-to-air missiles (SAM). These could be long to medium range or short-range systems including man-portable shoulder-fired ones. A big part of the ASC is the avoidance of fratricide. Most military operations are undertaken at short notice. The Air Force, Army and Navy have their own aerial platforms. There is thus a need for air space coordination between them especially in the TBA and in the sea when Air Force elements are launched in support of naval operations.

Layered Air Defence

Air is a complex medium. With many active players, the density of air operations, especially in the TBA has increased phenomenally. While we create a robust AD system, affording freedom of operations to own forces remains the key. It thus requires a great amount of command and control. Air defence of a vital asset or an area is normally built around a system of concentric layers. The outer layer will usually be handled by fighter aircraft with active electronically scanned array (AESA) radars and combinations of AD missiles supported by AEW&C. If an attacker is able to penetrate this layer, then the next layers would come from surface-to-air missiles. The area-defence missiles could have ranges in excess of 150 kilometres. Other shorter range missiles would have ranges around 30-50 kilometres. Finally, there will be the close-in-weapon-system (CIWS), the very short Range AD system (VSHORADS) missiles, the man-portable missiles and the radar controlled anti-aircraft guns firing several thousand rounds per minute.

Surface and Airborne Radars

Ground-based radars are an important element to manage both air threat and air space coordination. High and medium powered surveillance radars, tethered aerostat radar balloons, missile acquisition and guidance radars, tactical battlefield mobile radars and ship-based radars are all part of the ground sensor network. Radars such as Raytheon AN/MPQ-35 can detect high/medium-altitude threat for MIM-23 Hawk surface-to-air missile system. Northrop-Grumman AN/TPS-75 is a transportable three-dimensional air search radar. The Chinese ground based radars include the very powerful REWY-1 long range surveillance radar and YLC-18 medium range low-altitude 3D radar. The Russians have the 'Duga' series of over-the-horizon radars and Don-2N and Voronezh anti-ballistic missile radars. To cater for the stealth aircraft threat, Russia, China and Israel are developing very long-range L Band, UHF and VHF wavelength radars.

Air Traffic Control (ATC) radars and controllers perform a significant role in airspace management. India has a large number of ground radars such as indigenous Rajendra and Rohini, Elta medium powered radars (MPR), and GS100 low-level transportable radars (LLTR) developed jointly between BEL and Thales which are under induction. DRDO is developing 'Arudhra' MPR. Low level light weight radars (LLWRS) have been developed for the mountainous and high altitude regions. High-powered radars with a range of more than 500 km to replace THD-1955 are being identified. The IAF already has Israeli Rafael Aerostat radars with a range of 400 km. The DRDO is also working on the indigenous 'Akashdeep' aerostat.

The airborne early warning and control (AEW&C) system is an airborne radar picket system designed to detect aircraft, ships and vehicles at long ranges and perform command and control of battle-space in an air engagement by directing fighter and attack aircraft strikes. AEW&C is also used for surveillance and frequently perform battle management command and control (BMC2) functions similar to an ATC. It also allows the operators to detect and track targets and distinguish between friendly and hostile aircraft much farther away than a similar ground-based radar. Because of its mobility, it is much less vulnerable to counter-attack. AEW&C aircraft are used for both defensive and offensive air operations. The Northrop Grumman E-2 Hawkeye, the Russian A-50, the IAF's IL-76 based Phalcon, Chinese KJ-2000 and the Indian DRDO's Embraer-145 based Netra are some examples of AEW&C aircraft. Boeing's E-3 Sentry on Boeing-707, aircraft or more recently, on the Boeing 767 is another front line AWACS. There are a large number of helicopter AEW systems such as Sea King ASaC7, Augusta Westland EH-101A and Russian Kamov Ka-31.

Surface-to-Surface Missiles and Artillery

SSMs are designed to be launched from the ground or the sea and strike targets at long ranges. They have significant size and range. Ballistic missiles travel in a high trajectory. Even tactical missiles have ranges up to 300 km. A cruise missile could have range up to 2,500 km.

Many countries including the USA, Russia, China, Israel and India are developing anti-ballistic missiles (ABM) and conventional SAM systems. China has its own anti-ballistic missile & anti-satellite programme. The Israeli Arrow ABM system was very successful against the SCUD attacks. The Arrow 3 system under development will be capable of exo-atmosphere interception of ballistic missiles. The Iron Dome and David's Sling are designed to intercept tactical ballistic missiles as well as medium to long-range rockets and slower-flying cruise missiles up to 300 km way. India is developing its own Advanced Air Defence (AAD) missile system as a part of Programme AD.

Space Based Military and Civil Applications

Space has been used for positioning satellites using a variety of optical, infra-red (IR) and radar-based sensors for surveillance, accurate mapping, communications, data networking, global positioning system (GPS), etc. The dependence of space for ground operations has become phenomenal. For any war between major rivals it will be important to decapitate such systems to degrade the enemy war effort. The number of satellites is increasing every day. The day is not far when hypersonic airliners will transit through near space. Space tourism is already on the 'horizon.' The line dividing space and atmosphere is thinning and therefore the new reality is that space-based systems have a direct bearing on operations on the ground, adding a new dimension to ASC.

Space weapons can be categorised as those that attack targets in space (anti-satellite); or attack targets on ground from space; or attack targets transiting through space (anti-ballistic missile). The USA and the Soviet Union began developing anti-satellite weapons in the early 1960s. They were in the form of directed-energy lasers to decapitate; kamikaze satellites for hard-kill; and possible orbital nuclear weapons. Ground-based AD weapons have to cater for long range ballistic attacks transiting through space. The Russian ASAT research has reportedly resumed under President Putin to counter the renewed US strategic defence efforts post the ABM Treaty. The NASA space plane X-37, now with the US Department of Defence is akin to a space version of an UAS and its employability is evolving.

Research is on into directed energy weapons, including a nuclear-explosion powered X-ray laser. Many countries like the USA, Russia, China, India and Israel have anti-ballistic missile programmes, with exo-atmospheric interception capability. International space treaties limit or regulate positioning of weapons or conflicts in space. The air defence assets required to defend against a ballistic missile, SSM or a cruise missile may be similar.

Terrorists in the Air

The September 2011 coordinated air attacks over America by terrorists hijacking airliners and making suicide attacks against ground targets brought a new dimension to air threat. Any aerial platform or weapon falling into terrorist hands could thus have implications. Two immediate concerns are

9/11 - Changed the face of air threat forever

terrorists acquiring weapon-laden UAVs or surface-to-surface missiles. Motivated terrorists could train to join civil or military aviation and later indulge in suicide attacks. A terrorist has the advantage of choosing time and place of attack. While response to the threat would be conventional, better surveillance, policing, and prevention of weapons going into their hand is more important.

Connectivity and Cyber Threat

The entire operations today are based on network-centricity. Every platform is electronically talking to other platforms and sharing critical data. Commanders are taking operational decisions based on situational awareness (SA) created by networked sensor inputs. Each service has its own secure dedicated net. Also, there are inter-service networks for sharing common domain information. The entire aerospace management is based on connectivity. A major part of the cyber war will thus be to attack the surveillance and control systems of the enemy.

Any attack on the military networks can be critical for the outcome of the war. Cyber war doesn't require huge armies with attendant logistics and can be launched by a single operator with a simple computer. All networked civil utilities like water, electricity, banking, trade, transportation, etc., can be ground to a stop through a cyber attack which could have an impact as devastating as a nuclear bomb. Military communication and surveillance networks can be infected or blanked out with disastrous results.

Organisational or personal hardware and networks can be targeted. Malicious bugs can be inbuilt into computer systems or in microchips at the manufacturing stage. The time and place of attack can be chosen and therefore some also call it 'cyber-terrorism'. For any ground-based air defence network to succeed, it has to defend its various elements from cyber attacks.

Permeation of UAS

Permeation of UAS in the NAS is posing a new challenge in the area of air space control (ASC). The central issue contained in this is the fact that unmanned vehicles are growing at a fast pace in India and are likely to register an even higher growth in times to come. This is despite the fact that till very recently there was a total ban on the operation of UAS in the NAS If these vehicles were to cohabit the same finite air space where manned aircraft (civil and military) also operate, then there is a looming danger of a collision of the manned platforms with the unmanned machines which may lead to catastrophic destruction besides loss of innocent lives.

What even a perception of such a looming danger can result in was seen on 21 August 2017 when, at the Indira Gandhi International Airport (IGIA), Delhi, the pilot of an Air Asia Goa-Delhi flight sited an UAV-like object on its approach landing path. A panic situation ensued. All the three runways of the IGIA were closed down for nearly an hour, 20 flights got delayed and 20 police and CISF teams were hurled into action to investigate the incident.

Such a reaction was in a way justified considering the fact that at peak time, close to 70 flights operate each hour and IGIA has over 3,70,000 flights a year. Thousands and thousands of innocent lives can be lost by a potentially dreadful head-on collision between an aircraft and a uncontrolled, unscheduled and a 'loose-kite like' UAS. Understandably, such a closure at IGIA was not the first. Similar sightings of UAS were reported on 28 October 2015 and 10 December 2015 and had led to similar reactions.

As stated, the UAS in India are growing really fast. According to one estimate, India is amongst the top countries in terms of UAS imports and accounts for nearly 20 per cent of the total global imports of UAS. Besides imports alone, the 'Make in India' initiative of the Government which is taking root slowly but positively is also likely to push the domestic manufacturing of UASs in the coming years.

The growth of the UAS sector is driven by the demand of such platforms in multifarious functional areas, both in the defence as well as in the civilian domain. the former going much beyond the traditional reconnaissance surveillance and target acquisition (RSTA) role to electronic warfare (EW), deception operations, nuclear cloud surveillance, 24x7 watch by lighter than air (LTA) platforms to a host of other military applications leading to the mother of all—the teaming of the manned and the unmanned vehicles in joint operations. According to an independent assessment, the Indian armed forces may be looking at reducing the ratio of manned to unmanned platforms from the current 95 per cent to about 50 per cent by 2027.

In the civilian domain, it is the nearly limitless possibilities of the uses that this machine can be put that are the driver for its likely exponential proliferation in the NAS. From policing duties to traffic spotting to fisheries protection to pipeline survey to sports, film coverage, precision agriculture to aerial photography, disaster relief, emergency medical aid, to business deliveries to law enforcement—the limit is only a function of one's imagination. It is no wonder therefore, that besides the major public sector players National Aeronautical Laboratory (NAL), Hindustan Aeronautics Ltd. (HAL) and Bharat Electronics Ltd. (BEL) et al, a host of other major private industry leaders like Tatas and TAAL, besides many other small players, are gearing up to develop different types of UAS in India for which the technical expertise, as well as, the manufacturing capability is very much present in good measure.

Fresh Approach to Regulation

If all the above and more is rearing to go, then the problem of regulation of the UAS is the biggest challenge in the ASC domain today. The DGCA announced draft regulations (DRs) on the civil use of UAS in the NAS on

1 November 2017, After comments from the environment a final version is expected. A large part of which is on the lines as announced by the Federal Aviation Authority (FAA) in the USA. According to the DRs, the UAVs have been classified according to their weight, i.e., nano (250 gm or less), micro (250 gm-2 kg), mini (2 kg-25 kg), small (25-150 kg) and large (greater than 150 kg).

All UAS (refereed to as drones in the DRs) are proposed to be operated in visual line of sight during the day time only and below 200 ft. All commercial drones except those in the nano category and those operated by government agencies have to be registered by the DGCA as per International Civil Aviation Organisation (ICAO) regulations in the form of a unique identification number (UIN). The mini and higher categories will require an unmanned aircraft operator permit (UAOP). The DRs mandate that all the remote pilots must undergo requisite training except for those using drones in the nano and micro categories. All micro and higher category drones will have to be equipped with RFID/SIM with a return to home option and anti-collision lights.

The DRs also specify certain restricted areas for operation of drones. They cannot be operated within an area of 5 km from an airport or within permanent or temporary prohibited, restricted or danger areas as notified by the Aviation Authority of India (AAI) or without prior approval over densely populated areas or over or near an area affecting public safety or where emergency operations are underway, within 50 km of the international borders and beyond 500 metres (horizontal) into the sea along the coastline. Also, drones cannot be operated within a 5 km radius from Vijay Chowk in New Delhi and from a mobile platform such as a moving vehicle, ship or aircraft.

Civil Air Traffic

All non-military flying activity, commercial or private is encompassed in civil aviation. The International Civil Aviation Organization (ICAO) establishes common standards and recommends practices for safe civil operations, scheduled air operations including all passenger and cargo flights operating on regularly scheduled routes and general aviation (GA) including all other civil flights, private or commercial. Although scheduled air transport is a larger operation in terms of passenger numbers, GA is significantly large in number of flights.

Air Space Control—a key battle management function

Air travel is increasing exponentially. The USA alone has 350,000 flights a day. There are an estimated 500-odd civil airliners in India currently, and another 1,000 aircraft are on order. There is a huge network of civil radars and ATCs to coordinate and control safe operations. The IAF maintains liaison teams in all major airports. These teams facilitate civil-military flight coordination. They also monitor and coordinate flights during major operational air exercises and events like the Republic Day flypast. Another important task is to monitor possible rogue aircraft or major air route deviations having security implications. Like the military, the civil ATCs and radars are fully networked. The IAF has established a system of porting and super-imposing civil traffic data on to the IAF air picture. Till now, the civil flights followed pre-designated airways. Aircraft followed positive or procedural control within the airways or terminal areas. The future air navigation system (FANS) is an avionics system which provides direct satellite data-link communications between the pilot and the ATC. The communications include air traffic control clearances, pilot requests and position reporting. Wherever implemented, this will allow direct shortest distance flights between any two points. This will add another dimension to coordination with military flying and promulgation of no-fly zones aiding ASC in a big way.

Air Space Control in the TBA

General considerations when addressing airspace functions across the range of military operations include command and control (C2) and air traffic control (ATC). The airspace control order (ACO) provides the details of coordination measures for the air tasking cycle and includes fire support coordination measures (FSCM), air defence areas and air traffic areas along with other airspace information. Airspace de-confliction at the operational level normally occurs within the air operations centre (AOC). De-confliction at the tactical level is executed by elements of the airspace control system such as control and reporting centre (CRC), airborne warning and control system (AWACS), ATC, etc. The air defence functions of weapons control, surveillance and identification are inherent in the theatre air control system from the oversight and direction provided by the senior most air force officer controlling the air space.

The area air defence plan should provide detailed engagement procedures consistent with the airspace control plan (ACP). Airspace control and area air defence operations should be capable of functioning in a degraded C2 environment. Detailed engagement procedures and clear delegation of authority to subordinate commanders are key to its success. Integrated air and missile defence (IAMD) capabilities and overlapping operations to defend national and military assets, yet enabling freedom of action by negating an adversary's ability to achieve adverse effects from their air and missile capabilities is of essence.

The air component commander must ensure that items on the surface commander's defended asset list and critical asset list are protected from attack while simultaneously minimising friendly fire incident risk between friendly forces. Airspace coordinating measures are employed to facilitate efficient use of airspace to accomplish air operations and fires and simultaneously provide safeguards for friendly forces. All airspace users should coordinate with the appropriate airspace control. The volume of air traffic demands careful coordination to limit potential conflict among mission essential aircraft and other airspace users.

Fire support coordination measures are employed by commanders to facilitate the rapid engagement of targets and simultaneously provide safeguards for friendly forces. It is important to define the theatre-specific area map reference system in use so that air and surface elements remain geographically on the same grid. Detailed analyses of joint network and joint infrastructure requirements are crucial in order to enable system integration across components. Planners should integrate surface-to-air weapons and air defence aircraft for maximum effectiveness. Proper coordination with civil air operations is especially important during transitions into or out of wartime status or during non-wartime periods of heightened tension. Preventing collision between aircraft and with other airborne objects is a task. Additionally, allowing maximum use of navigable airspace by civil aircraft are necessary.

TBA - Indian Scenario

TBA is the area around which surface forces are engaged in battle. In the TBA, the enemy air tries to engage our surface forces. Similarly the IAF will support the Indian Army through air operations. There will be many joint or special operations. Between the Indian Armed Forces, the domains are clearly demarcated. The Army manages the surface coordination, the Navy manages the maritime picture and the Air Force coordinates the ASC. The air defence of the nation is the IAF's responsibility. The air defence of the Army and Navy's integral assets is their own responsibility. The big situational air picture is created by the IAF using its own, civil and other services radars. Such a picture is made available at the tactical air control (TAC) level to the Army and at the Maritime Element of the Air Force (MEAF) level to the Navy. The air defence clearance for all air movement is given by the IAF. Very low flying army air assets within a small bubble of air space do not require any clearance. Similarly inter-ship helicopter naval flights are managed by the Navy. All flights within the air defence identification zone (ADIZ) require IAF air defence clearance. Naval flights beyond the ADIZ are managed by the Navy. IAF attack and support aircraft flying in support of the Indian Navy beyond the ADIZ are coordinated by the Indian Navy. Such coordination is very necessary to avoid fratricide. Any hold-fire order passed by the IAF would be for short durations over a small geographical area so that full-scale operations of the Army/Navy are not hampered. Low-level routing of IAF aircraft through the TBA is normally through points in joint knowledge. The IAF aircrew acting as forward air controllers (FAC) also support ASC at the tactical level. There is interface between the IAF and the Army at Corps HQ and Command HQ levels to iron out day-to-day issues and jointly monitor the progress of battle. Similarly, air elements operate with the Indian Navy.

To carry out the above tasks, detailed arrangements are in place at successive levels of command (Corps/Division/Brigade etc). The apex control of ASC is with the Air Force being the largest user of air space. The instructions for allowing or denying the use of air space to a user are laid out both as standing instructions (height bands, time slots, fly/no fly zones and more), as well as, in the form of dynamic and instant instructions applicable to a user at a point in time. The aim behind all the ASC activity is to allow maximum freedom of operation to every potential user with minimum restrictions.

Also, the Air Force is in the process of aligning the ASC function through its Integrated Air Command and Control System (IACCS) designed for controlling and monitoring air operations by the Air Force at the strategic (Service HQ), operational (Commands) and tactical (Division) level cutting across Service boundaries. The ASC organisation also takes in its fold the civil aviation with detailed and institutionalised tie up between the Air Force and the Directorate of Civil Aviation (DGCA).

Military Civil Coordination

Any threat to India will require an integrated approach of all national air assets. The Indian Armed Forces will requisition airliners and cargo planes for inter-theatre movement. Similarly, civil helicopters will be used for communication and air ambulance duties. Expressways will be used for operations in emergencies. A key element will be civil radars and ATC networks. Networking, liaison and peacetime training will make the process more efficient. Military aircraft would be accorded direct routing priority. There will be height band restrictions on civil traffic during operations.

Dual-Use Airfields

A large number of military airfields are being used for civil traffic. Conversely, there are a few civil airfields being used for military operations. Dual-use airfields have typical operational peculiarities. Most fighter aircraft require arrestor barriers or hooks. There are peculiar security issues for military airfields. Also, many airbases will have fully armed aircraft on operational readiness platforms (ORP) for take-off at short notice. Procedures for approach and landing of a battle-damaged aircraft are considerably different. Civil aprons may be used for dispersal of IAF assets. ASC has to factor in all these peculiarities.

Challenges

Manned aircraft and UAS are going to operate in the same airspace and at the same time and avoiding collision between them will be a major issue and a major challenge. They will have intelligent systems on board and will exchange information with air traffic and fighter controllers over high-speed digital data links. On-board collision avoidance and advanced traffic display systems will greatly improve the situational awareness of pilots. The cockpit will be 'information rich.' As this future on-board intelligent environment evolves, it will be critical that we ensure the integrity and security of data. There will be a need to sift the huge data for relevant dissemination, display and use. Human-computer interface will be crucial. For emerging powers like China and India, the big challenge is to get all the agencies and systems to switch to newer technologies together. The TBA will have very accurate lethal weapons. The air traffic and projectile density in the TBA will continue to increase. Humans will have to bank more and more on machine intelligence. The airspace in future will be 'dynamic' and following point-to-point navigation.

Artificial Intelligence in Air Space Management

The development of computer systems to be able to perform tasks normally requiring human intelligence such as visual perception, speech recognition, decision-making, and translation between languages, i.e., artificial intelligence (AI) has great scope for ASC Intelligent machine systems can interpret complex data; perceive the environment and take appropriate action using learning and problem-solving techniques.

AI processes include perception, reasoning, knowledge, planning, learning, statistical analysis, computation and, finally, manipulated output. It has evolved using expertise in fields like computer sciences, mathematics, psychology and neuroscience, among many others. AI applications already exist in industrial machines, automotive industry, surgery and aviation, among others. Another school of thought is to use the phrase 'extended intelligence' to signify how AI is used to augment human decision-making rather than replace it.

The London Air Terminal Control operations are already using AI for airspace infringement monitoring. Possible links between safety events and variables such as high traffic volume, airport runway direction, weather conditions and more are being predicted to avoid potentially avoidable events, though there is still some way to go to dispel the perception of AI as coming from a faceless 'black box'. The trust would flow from accuracy of the output and through the extra value we add by expert human analysis of the data and optimise decision-making to help meet future regulatory targets. It will help Go-No-Go decisions. Possible control strategies are identified, ranging from visual and electronic collision avoidance to de-conflict flight paths. AI will relieve the air traffic controller from the current chronic fatigue. The Defence Advanced Research Projects Agency (DARPA) has been working with Lockheed Martin's Advanced Technology Laboratories to develop AI technology that keeps airspace operating safe despite UAS and other airborne weapons

Such is the dynamic world of air space control where the bottom line challenge is to provide maximum freedom of operations to the multifarious users of air space with minimum restrictions, while at the same time, ensuring safety of air traffic and keeping fratricide at bay. By any standards it is a huge, huge challenge!

CHAPTER 5

The Unmanned Revolution: Attacker's Delight, Defender's Nightmare

Lt Gen (Dr) V K Saxena, (Retd), PVSM, AVSM, VSM

This chapter is based on the premise that with the arrival of unmanned machines, the battlefield has never been the same. In that, while on one hand, the unmanned aerial systems (UAS) are becoming an 'attacker's delight' as strong and powerful tools of combat, on the other, these are proving to be the 'defender's nightmares' owing to the multifarious challenges involved in countering the air threat posed by them.

While this eternal duel between the prosecutors of the air threat and defenders is getting fuelled by niche technologies on both sides, the latest buzzword in the unmanned world today is MUM-T, which stands for Manned and Unmanned Teaming. MUM-T is that win-win situation in the combat domain where the intellect, grit, instant decision-making capability and the hunch of the combat pilot are being combined with the precision, endurance and the all-weather, all-terrain capability of the unmanned machines. The gains are disproportionate and multiplicative, but all these not without their share of challenges; some operational, some ethical.

Flowing from the above, this important chapter has been attempted along the following three verticals:

Part 1. UAS—An attacker's delight

Part 2. UAS—A defender's nightmare
Part 3. MuM-T—The latest buzzword in UAS

PART 1: UAS—AN ATTACKER'S DELIGHT

UAS Has Many Combat Virtues...

There are many virtues, which make the UAS a versatile threat vehicle. There are no crew fatalities, no crew fatigue and no training deficit on board, given the expertise of the pilot at the ground control station (GCS). As the technology has shown, the weapon loads lifted by UAS have become comparable to manned aircraft. Sample the weapon load of the Predator (B) - 'Reaper', the largest and the most powerful variant of Predator UAS. It can carry eight laser-guided AGM114 Hellfire missiles besides two GBU 12 Paveway II laser guided bombs complete with its infrared sensor, TV camera, image intensifier night camera, a laser designator and a laser illuminator. Hellfire missiles could be exchanged with AIM 9 Sidewinder missiles or with another PGM. A machine with such a load and an endurance of 42 hours (with 1000 lb) or 14 hours (when fully loaded) is a formidable platform indeed, comparable to a manned aircraft.

Predator B 'Reaper' UAS

UAS also score over manned platforms in terms of having no restrictions to operate in hazardous/contaminated environments. Besides this, their range, reach and precision combined with tremendous endurance (Global Hawk UAS holds the world endurance record of 30 hours, 24 minutes and 01 second at an absolute altitude of 65,381 ft) gives them added advantage.[1] The possible mission spectrum of UAS is indeed expansive. Pre-ingress route reconnaissance, nuclear biological and chemical (NBC) detection/ sampling, battle damage assessment (BDA), electronic warfare (EW), signal and communication intelligence (SIGINT and COMINT), surveillance, reconnaissance, battle management command and control (BMC2), precision attacks on chosen targets of interest (hunter killer missions), directing gun fire, etc., to name a few.

Global Hawk UAS

The strength of these machines lies in their open architecture, flexible mission payloads, video-on-demand capability, instant connect (typical time for a control message to UAS on a satellite downlink is 1-2 seconds). Capability for dynamic in-flight re-tasking, high manoeuvre and near real-time high data transfer rates are some other strengths.

In the light of the growing clout of UAS, there is much talk of their comparison with manned platforms by comparing the virtues of one against the other. This is a futile exercise since the UAS are not a complete replacement of the combat aircraft as the two are likely to co-exist in a non-mutually exclusive domain in a net-centric battlefield. The idea of co-existence actually stems from the respective vulnerabilities and limitations in both the manned and unmanned platforms respectively and how both can largely address each other's sub-optimalities in joint operation.

UAS—Not without Vulnerabilities

While any number of combat virtues may be counted in favour of UAS, these machines are also limited on many counts. Most prominent is their vulnerability to minimal air defence. For instance, while the Predator UAS succeeded in striking down several Iraqi radar units during operation Iraqi Freedom, these quickly became targets for Iraqi air defence. The Iraqi Air Force actually shot down three Predators with relative ease. Even in areas where there is no strong air defence cover, UAS are vulnerable to ground fire from small arms and rockets normally held by ground forces or even with terrorists. For example, even in the tribal region of Pakistan, where there is virtually no air defence, the Taliban has claimed to have shot down several CIA drones over South Waziristan.[2]

Iran claims that its SU-25 Frogfoot warplane had shot down a US Predator drone. Even when not facing enemy fire, Predator statistics show that it has crashed due to mechanical error 43 times per 100,000 flying hours whereas a typical manned aircraft crash data shows 2 per 100,000 flying hours. Lack of military presence on ground also limits the capability of the drones to assist in acquiring critical intelligence. Also, the so-called Hunter killer operation can only eliminate terrorists but fail to get intelligence as 'dead men tell no tales.' Besides this, there is a huge issue of collateral damage in such missions by way of casualties of 'unintended targets.'

UAS driven by the way point navigation provided by the global positioning system (GPS) are perennially vulnerable to jamming for which jammers are freely available on the internet. The latest in this field is the 'spoofing' of a GPS receiver on board a UAS, thus taking control of the machine much in the same manner as hijacking a plane. The UAS thus taken under control can be made to do what the spoofer wants, viz., to crash on a safe area or be retrieved safely for study and analysis and, worse still, to be used as a missile on suicide mode crashing on desired targets much like the 09/11 of attack.[3] Also, a UAS cannot take a decision in the situation of ambiguity, nor can it take on opportunity targets that come up suddenly. It is a sitting duck for strong air defence and is devoid of pilot's 'hunch' and IQ; a reason enough for MUM-T.

Cutting-edge Technologies Empowering the UAS

UAS is an arena of hot technological research. Many new ideas are becoming realities. Some of these include miniaturisation of electronics, and overall structures., qualitative improvement in sensors and building reliable and jamresistant data links. Continuous research and trials are also taking place in the operation of UAS from aircraft carriers.[4]

In another trend, the leading combat aircraft manufacturers of the world are working towards launching the unmanned versions of their strike fighters together or shortly after the launch of the manned version. In this context, there is open source information regarding M/s Lockheed Martin planning to launch the unmanned version of their latest F-35 Lightning II Stealthy joint strike fighter soon after the launch of its manned version.

Besides the many combat virtues of the F-35, the unmanned version can assume riskier mission profiles. Since no pilot weight is to carried, the aircraft can carry extra combat load by mounting external pylons. By doing that the useful payload can be increased from 5,200 to 18,000 lb albeit with reduced stealth).[5] Also, the unmanned version is reported to have the capability of STOVL (short take off and vertical landing), making it suitable for operation from aircraft carriers.

MQ 8 Fire Scout VTOL UAS

Vertical/short take off and landing (VTOL/STOL) is another emerging concept. Northrop Grumman's MQ-8 Fire Scout is an example. This STOVL platform is designed for reconnaissance, situational awareness and precision targeting support for US ground, air and sea forces. It has an endurance of three hours, a range of 100 nautical miles with modular mission payloads to include Hellfire missiles, Viper strike laser guided glide weapons and has pods carrying the advanced precision kill weapon systems. Laser target designation, gimbled Electro optical/IR payloads, day/night cameras, coupled with laser range finders provide it an all-weather operational capability.[6]

Another emerging concept in UAS is to achieve full vehicle autonomy. In fact, air vehicle autonomy is a recently emerging field. It aims to reduce the dependence of an UAS from ground control infrastructure making it a self-contained and autonomous war machine. Achieving such an end-state will call for realisation of many a cutting edge technologies. Firstly, the sensor-fusion technology which must combine information from different sources for single-window usage on-board the vehicle. Motion planning and trajectory generation technology which must determine an optimal path for the vehicle either to execute optimal control manoeuvre or to stick to the path defined and the artificial intelligence tools which provide a degree of autonomy in, decision making and precision kill capability based on dynamic parameters available at the mission end (face/target signature recognition and more). The idea is to teach the machines to be 'smart' and act more like humans. In sum, autonomy is the end state that is being achieved by using multiple technologies such as artificial intelligence, expert systems, neural networks, machine learning, natural language processing etc.

Many other exciting new technologies are entering the field of UAS in almost the same time sequence as for combat aircraft. One such technology is morphing, i.e., changing the configuration of the vehicle itself to maximise performance in different flying conditions. Morphing aircraft are multi-role aircraft that can change their external shape substantially to adapt to a changed mission environment during flight.[7] Variable sweep wings, landing and takeoff flaps, retractable landing gear, variable incidence manoeuvre, etc., are some established morphing forms.

Morphing in UAS is becoming increasingly visible with possible areas being wing/fuselage/engine/tail. All the possible morphing techniques, viz., folding/hiding/telescoping/expanding/contracting/coupling/de-coupling, etc., are being applied to bring changes in aspect ratio, wing area, wing twist and more. This provides variable performance in different modes like longer loiter time with efficient dash speeds, high responsiveness for time critical deployments, high agility to attack fast-moving ground targets and long persistence to dominate large operational areas.

Besides the conventional morphing configurations, the next generation morphing designs include distributed actuators to increase survivability and decrease weight, stretching or sliding skins that meet the changing requirements of symmetric morphed wings and/or conformal flaps for flight control. An innovative morphing design by M/s Lockheed Martin hides a substantial portion of the wing area during low altitude, transonic dash portion of the mission. Another morphing design by M/s Raytheon is based on telescoping, which is ideal for controlling aerodynamic drag. The leap of technology indicates that morphing technologies are going to find more and more usage in the UAS of today and tomorrow.

Orinthropters are another emerging variation of UAS. These can fly by flapping their wings just like birds. M/s Aero Vironment of USA is a lead player in this field. Orinthropters supposedly started their existential journey from the fourth century BC, where, in the epic *Ramayana*, there is a mention of 'Pushpak-Vimana,' the orinthropter of the gods; this innovative design has indeed come of age.

Unmanned orinthropters have many practical applications. They can carry out aerial reconnaissance without alerting enemy forces. The designers hope to eliminate the motors and gears of the current design by more closely imitating the animal flight muscles. Research is already in progress for developing a reciprocative animal muscle for use in such UAS. As opposed to a jet engine, which produces a relatively narrow stream of fast air, an orinthropter displaces a comparatively large mass of air at slow speed, thus enhancing efficiency in flight. The lift and thrust generated by the flapping wings can be set to a zero angle of attack in its upstroke. This allows the rinthropter to pass easily through the air, thus reducing drag.[8, 9]

M/s Aero Vironment's Humming Bird

Nanotechnology has also impacted UAS development in a big way. M/s Aero Vironment's Hummingbird is only 7.5

centimetres/10 grams with a flying speed in excess of 15 miles per hour with precision hover and live video streaming capability. This surveillance UAS made it to the front cover of the Invention Issue of TIME Magazine which featured the same as one of the 50 new inventions of its time.[10]

A similar nano UAS from the same firm is called 'Black Widow.' This micro UAS weighs less than 50 grams. It has a flight speed of 200 feet/second, endurance of 30 minutes and a hover endurance of 8-10 minutes. It has the capability of controlled transition from outdoor to indoor and precision flying in urban settings. There are several other UAS in the light/very light range. The RQ-11 'Raven' of M/s Aero Vironment is a small hand-launched UAS which is back-packable and weighs only 1.9 kilograms. It has a range of 10 kilometres and an endurance of 60-90 minutes. The payload is an electro optical (EO) camera and an infrared (IR) camera with a capability of real-time video imagery, especially for the short tactical requirement of 'over the hill/around the corner'.[11] Lockheed Martin's 'Desert Hawk' UAS is only 3.2 kilograms while Aero Vironment's 'Switch Blade' is only 2.5 kilograms.

Research is also on in the field of pseudolites (a contraction of the term pseudo-satellite). It is something that is not a satellite but performs the functions commonly in the domain of satellites. These are often small trans-receivers which are used to create local ground-based GPS alternatives.[12]

The SWARM (smart war-fighting array of reconfigurable modules) is another manifestation of nano-technology application in the UAS that is fast emerging. In this, many a small autonomous UAS are made to fly in a swarm without using the ground control station and programmed to deliver continuous co-ordinated information to enhance situational awareness. Individual vehicles in the swarm are so programmed as to communicate and co-ordinate actions of each other and one another to conduct autonomous searches. The data collected is aggregated and relayed back in real time.[13] Small-scale UAS functioning as SWARM rely upon each other in a similar way as migrating birds. They can collect data precisely, navigate through complex and difficult terrain and are continuously kept updated about their surroundings, status and their inter-se position with respect to the swarm. This is achieved by using evolutionary algorithms.

As a result of the above networking algorithms, if out of a large swarm formation, a few UAS land due to malfunction or have other problems such as engine failure, etc., the control system ensures that the other UAS become aware of this and form new formations that allows the rest of the swarm to collect the data which the malfunctioning UAS was/were supposed to be collecting. This allows the swarm to function as a 'live mind' of sorts, making decisions in the autonomous mode.

M/s Boeing is one of the several companies that is developing swarm-based technology. The same is based heavily on the communication systems used by animal swarms. Suitable algorithms are being developed for performing a variety of tasks such as searching a designated area, mapping it, creating waypoints and generating information which can be relayed to the network in near real time.

A Symbolic SWARM Attack

M/s Ecole Polytechnique Federale de Lausanine in Switzerland has developed a 'Swarm Software' (called SMAVNET) for use in disaster situations. SMAVNET or 'swarming micro air vehicles network' (each vehicle weighing 420 grams and having a wing span of 80-centimetres) is being designed as a communication system relay in disaster situations to aid rescuers. The software uses a gyroscope along with pressure sensors to make decisions as to which flight path is better than the other. Their communication abilities mimic that of 'ants' which use pheromones to remember the path they take from their nest to food. The SWAMNET uses a similar principle to allow the UASs to paint a path based on remembrance so that it can have a constant communication path to the home base for sending information back.[14]

The day is not far when we will see swarms of operationalised UAS, each small vehicle lifted and propelled by rotors (four rotors making the vehicle a quadcopter) and moving like a flock, much like storks or wasps or ants. Each small vehicle will be tied to the other in the soft mode through yet evolving algorithms which will allow them to operate as a 'team.' The future of intelligence, surveillance, target acquisition and reconnaissance (ISTAR) missions is here!

In the fascinating and ongoing journey of the evolution of SWARM UAS, 5 January 2018 became a milestone since it was on this day that the Russian air base at Khmeimim and its Tartus naval facility in West Syria were struck by a SWARM of home-made drones. This is the first time that such a coordinated assault had been reported in military action.[15] Russia claimed that it had identified the launch location from where the SWARM of armed drones began their attack.[16] Details revealed indicate that ten drones rigged with explosive devices attacked the air base while three others attacked the naval base. Though the attackers remained unknown, the Russians blamed Syrian rebels backed by Turkey for the terrorist attack. Be that as it may, SWARM technology is now battle-baptised and the sky is not the limit for the flight of this emerging technology.

Stealth technologies are also impacting the modern-day UAS in a very visible manner. Many leading nations in the forefront of UAS capabilities are adopting stealth techniques as a measure to enhance survivability. Lockheed Martin's RQ 170 Sentinel is an example of a stealth UAS. This tailless flying wing is a stealthy machine which has pods, presumably for sensors or SATCOM having been built into the upper surface of each wing.

As the time unfolds, fourth generation UASs will come on the scene and will be more robotic than merely unmanned. They would be characterised by enhanced autonomy thereby removing the need for constant controlling (total vehicle autonomy). Artificial intelligence (AI) would take over more and more of routine tasks including decision-making at some point in time. Automatic returning capability would be further enhanced and so will be the craft's ability to assess situations and decide on courses of actions by superior algorithms. Such machines will be more manoeuvrable and thus capable of air-to-air combat and multiple target strike operations with more precision and shall have an ability to self-defend.

M/s Elistair unveils next generation tethered UAVs

The Indian UAS Market

UAS have been in use in India ever since 1990. However, it is in the last few years that the UAS market has seen a surge. The UAS in use in India are the Israeli Searcher Mk 1 and Mk II, Heron and the DRDO's Rustom. Their mission spectrum has been operations at the borders to include intelligence, surveillance and reconnaissance (ISR), traffic control, rescue operations, communications, etc.

According to open sources, the future requirement for UAVs in the period 2014-27 is likely to be 60-100 for the Indian Navy, 300-500 for the Indian Air Force and 1500-2000 for the Indian Army. With this projection in force, the UAS market size in India is likely to grow from USD 1.6 bn in 2014-15 to about USD 4 bn in 2027.[17]

The future is also like to see a balanced mix of unarmed and combat UAS initially with conventional weapons and later with soft kill weapons.

Keeping in mind the high demand of UAS in India in the future, many a start-ups, as well as, established industry players have entered the UAS market bringing with them a variety of capabilities in the mini/micro/ conventional UAS sphere. Some of these include Alpha Design Technologies, Ashok Leyland, Aurora Integrated Systems, Dynamic Technologies, IDEA Forge and others. Many of them have obtained industrial licences and have tied up with international OEMs. In this, two US-based OEMs, namely, M/s Aero Vironment and Boeing and two Israeli OEMs, namely, Elbit Systems and Aeronautics have shown interest in the RFIs issued.

Elbit Systems has tied up with Hindustan Aeronautics Limited (HAL) and Aero Vironment has announced a UAS partnership with Dynamic Technologies. Both the companies are currently working towards developing the Cheel UAS as a 'path finder project' under the aegis of the US-India Defence Technology and Trade Initiative (DTTI).[18]

DRDO's Rustom II

Alongside this, the DRDO has also been reporting success with its Rustom UAS. It was reported in February 2018 that it had successfully completed the test flight of its Rustom II UAS at the Aeronautical Test Range at Chitradurga, Karnataka. Rustom II is a part of the Rustom line of UAS that includes, Rustom-I, Rustom-II and Rustom-H. Rustom-I is a tactical UAS with an endurance of 12 hours. Rustom-II is a medium altitude long endurance (MALE) UAS with an altitude capability of 22,000 ft and an endurance of 20 hours. Rustom-H is a larger UAS with an endurance of over 24 hours.[19]

PART 2: UAS—A DEFENDER'S NIGHTMARE

Bringing Down the UAS: A Formidable Challenge

Turning the coin over and looking from the point of view of air defence warriors, it is a fact that UAS is as much an attacker's delight as it is a defender's nightmare. The most prominent talk in the defender's domain is that of the growing imbalance between the total cost of a UAS vis-à-vis the cost of the SAM/manned/ground-based kill system that may be deployed to kill it. Since the latter is far, far higher than the former, the major concern today, is to reduce the 'cost' of killing the UAS by using innovative technologies, preferably in the soft-kill mode.

Challenges Galore in the Path of Bringing Down a UAS

Bringing down a UAS throws up multiple challenges. The first is the 'detection challenge' which means that it is difficult for the radars to detect the UAS due to their low radar cross section (RCS), small size, low acoustic signature and low infra red (IR) signatures. Though the Predator/Reaper type of UAS flying at about 100 miles per hour at low altitudes may be a child's play to shoot down for a manned jet fighter, however, if a drone flies higher, it will pose a problem for electronic detection due to its lower RCS. Even a Predator is undetectable beyond 3,000 feet and inaudible beyond 1,000 feet. The smaller drones have their survivability (implied high degree of difficulty in detection) built into their miniature size or stealthy features. Both these manifest as low RCS signature.

Dealing with Detection Challenge

Defenders are busy researching new and innovative methods to detect small UAS. Cutting edge research has shown that phased array radar technology coupled with the tremendous growth in radar computing and processing power has the capability to pick up small manoeuvrable UAS using the technique of electronic beam switching.

UAS possess lower speeds in comparison to aircraft which aids in radar detection. In this scenario, pulse Doppler radars with moving target identification (MTI) facility prove to be quite effective. The optimal radar frequency bands include S band, X band or a combined S/X band. For detection of small, un-cooperative, low RCS UAS in high clutter environment, the trend is to go for higher speeds of scanning incorporating multiple solutions on the same platform, viz., pure RF/radar solution, electro-optical solution or even visual warning, as the last resort.[20]

Another vulnerability of UAS is its electro-magnetic uplink and downlink. The same is being used by passive sensors to perform the function of alerting and cueing. Equipment in this category include IR alerters and acoustic alerters. A typical passive IR detection system is a two-tier system to include a wide-area, panoramic search sensor platform with passive IR sensors and a tracking and verifying sensor platform with a high resolution IR sensor and a laser range finder (LRF). Once a tentative track is established by wide-area search sensors, the verification sensor's line of sight is directed instantaneously to the track co-ordinates indicated by the search sensors with the help of rapidly steerable mirror gimbals. A closed-loop control using the viewer's IR image links the LRF's line of sight on to the target for precise laser ranging. The 'target in range' is determined by means of computed range measurement. Such silent-mode air surveillance systems are normally employed in conjunction with other active systems which aim to have the incoming threat on a closing-in-track with the silent passive system focused on the end game. IR sensor-based detection systems are also ideal for areas in the 'radar shadow' or as a gap-filler in certain geographical areas.[21]

A number of ESM-based solutions are also useful in detecting the UAS through recognition of their RF/EM signatures. Though the ESM-based sensors have been around since World War II, a host of new technologies have entered this field lately to improve the probability and accuracy of detection. The traditional frequency range is between 1-2 GHz to 18 GHz. Some technologies being put to use include 'Instantaneous Frequency Measurement' and higher speeds in sampling and digital processing of signals.[22]

Acoustic sensors have the advantage of not being restricted to the line-of-sight operations. Besides, these can have all-weather operational capability. They can also deploy search-while-track capability and can provide useful track information about UAS. Unlike their electro-magnetic or electro-optic counterparts, acoustic sensors are capable of searching all frequencies and all angles allowing for a wide open range capability. Additionally, heat and electromagnetic radiation-seeking sensors are easily fooled through countermeasures like dispensing flares or jamming while, to avoid an acoustic sensor, the aerial vehicle must hide its acoustic signature. This is not easy, as many a UAS cannot operate without generating a certain (albeit low) acoustic signature. Acoustic sensors have the further advantage of operation in cloudy or overcast battle environments undeterred by smoke blanketing. On the flip side, such sensors do not have long-range capability. Besides, these are also affected by atmospheric conditions. Also, the presence of mean temperature profiles can cause sound waves to refract, thus reducing the effectiveness of such sensors.[23]

The Kill Challenge

Once detected, UAS are soft targets for conventional hard kill systems. Conventional fair-weather/all-weather very short range air defence (VSHORAD) systems consisting of guns and man-portable/short-range SAMs (such as Star Streak, Igla 1M, Strela 10M, OSA-AK, etc.), have the capability to kill a UAS provided it has been detected either visually or electronically, by their surveillance/tracking/missiles guidance radars.

During the Kosovo War, machine-gun fire from helicopters was effectively used to shoot down UAS in the visual domain. Yugoslav forces also brought down a few Pioneer UAS using ground-based fire. In the context of UAS kills, experts, however, count gains against losses in terms of the number of 'pilot lives saved' and the 'cost of doing business'.[24] In yet another kill option, defender's UAS are designed to be pitched against adversary's UAS in a kinetic-kill mode wherein a mini/micro UAS can be used to hit the target UAS. In this, the challenges of initial acquisition, tracking and providing continued guidance to own UAS need to be surmounted. Active seeker/fire and forget, laser designation, riding on the guidance beam of a mother system, and more, are options under active implementation.

Besides the VSHORAD system, designers of several SRSAMs have claimed that their weapon systems are capable to take on the UAS. The Israeli Armament Industries (IAI) claims that their Barak SRSAM is capable to take on UAS besides aircraft and anti-ship missiles. While Barak-1 has an operational range of 10-12 kilometres, its higher versions (Barak 8) goes up to longer ranges 70/100/120 kilometres. Similarly, the Spyder (surface to air python and derby) anti-aircraft missile system developed by M/s Rafael of Israel is also claimed to be effective against unmanned systems.

Open media also report about engagement of UASs using manned aircraft. In one such event, a manned US aircraft was launched to pro-actively kill a Reaper UAS over Afghanistan when its ground control station lost control over the drone.[25] There is a report of Israeli manned aircraft shooting down a UAS by using an air-to-air missile (AAM). The UAS was alleged to be an Iranian UAS launched from Lebanon and tasked to carry out reconnaissance deep inside Israeli airspace.[26] Iran probably used its SU-25 fighter jets to fire on a US Predator UAS in November 2012.[27]

While the hard kill options for killing the UAS using SAM system or worse, the manned aircraft may appear to be straight, it is obviously badly out of balance in terms of cost considerations between the arsenal being used and the target being addressed. Besides, such hard kill measures can address individual targets or, at best, a few; what about SWARMS? For these threats, such options may neither be viable nor effective.

The most potent option to take on a UAS is the soft-kill option. This kill method is anchored on debilitating the real-time data link between the UAS and its GCS. The UAS's ground-to-air or air-to-ground data link signal to its GCS can either be jammed (which in all probability may result in the UAS abandoning its mission and returning to base) or be hacked and exploited, in which case the UAS's control can be taken over from the adversary and made compliant to the defender's commands.

In the above context, Iran claimed that its electronic warfare unit downed a US RQ 170 Sentinel UAS that had violated Iranian airspace. The photographs of the downed UAS showed the aircraft relatively undamaged except for minor damage on its left wing. This ruled out the possibility of a crash/engine/navigational malfunction. Experts feel that either it was a cyber/electronic warfare attack system (indicating a soft kill) that brought the system down or it was a glitch in the command and control system.[28]

Essentially, the soft killing of UAS involves attacking its communication links. Normally, UAS use a line-of-sight radio link in the military C band 500-1000 MHz or satellite communication in the KU band between10.95-14.5 GHz (uplink band to satellite is normally13.75-14.5GHz and the downlink from the satellite is normally 10.95-12.75 GHz). If the UAS's communications to its GCS are jammed, then the GCS operator becomes blind and the UAS flies around till it crashes or is out of fuel. In order to block the safe return route of UAS to GCS, both the above communication links must be jammed.[29]

There was a report in the open media from the Russian aircraft industry site, Aura Port, that efforts are on to incorporate ELINT-based UAS killing capability in a ground based short range AD weapon system. For this, the existing radars of the ground based AD weapon systems (GBADWS) are being revamped with ELINT stations along with opto-electronic sensors, thus making such weapons capable of detecting the EM signature of the UAS. The soft kill is achieved by jamming the communication data links of the UAS. There are also reports of Britain developing a directed energy (DE) weapon package applicable for use against UAS. The kill energy was to be derived on the higher power microwave route. Specific details about the weapon system are not known.

A conceptual laser kill vehicle

Laser-based killing of UAS is already an established option. Boeing mounted a 1 KW laser system on its established Avenger platform and showed its capability to shoot down a handful of small UAS. In the next phase of this weapon system, it installed a 10 kilowatt solid-state laser on a higher energy laser mobile demonstrator to take on various threat representative targets which include UAS.[30] M/s Raytheon has already demonstrated its capability wherein a 50-kilowatt solid-state laser had a UAS in flames. The manufacturers are developing a low-cost directed energy laser system to complement the close-in-kinetic energy systems.[31] Laser area defence weapons are already a reality. They promise to provide an effective kill solution to short-range threats like rockets, missiles, UAS and such other targets. The kill speed is the speed of light and the magazine is nearly 'unlimited.'

Kinetic kill vehicles to take on UAS types of threats are old wine. 'Cougar' is a kinetic kill solution manufactured by M/s QinetiQ of UK as a cost-effective counter to low cost tactical UAS. In this system, a low cost un-cooled long wave IR seeker is used to acquire the target and guide an interceptor on a collision course. The Cougar interceptor is designed to approach a UAS threat with closing speed that will be sufficient to cause catastrophic structural damage to the target giving an observable hard kill. The target UAS's low signatures in all wavebands necessitates mid-course guidance to place the Cougar interceptor in the collision course from where its on-board seeker could acquire and home on to the target.

Peregrine Eagle is another air-launched decoy/drone system employed as an anti-UAS system in the kinetic energy kill domain having a range of 800 miles with a service ceiling of 40,000 feet. It employs high-power microwave and is an electronic counter to UAS control.

Integrated Drone kill Systems - A New Trend Catches Up

Normally the drone kill systems till very recently, used to feature one type of kill solution on one weapon configuration, for instance, kinetic kill or laser kill or electronic kill and so on. The trend which is now catching up is to provide multiple sensor and multiple drone-kill solutions on a single weapon configuration. These are called Integrated Drone-Kill Weapons.

Several such class of weapons were on display in the recently concluded international defence and security industry trade fair in France which goes by the name of Eurosatory (11-15 June 2018). In this trade fare, M/s Skylock Anti-drone Systems, a part of Avnon group (an Israeli privately held company specialising in turnkey solutions for defence and security markets) showcased their system called Skylock Anti-drone system.

Skylock Anti-drone system at display in Eurosatory 2018

This system provides two sensor solutions, namely the radar solution, as well as, an electro optical (EO) mated with each other in a complementary mode for the detection of drones and two kill solutions, namely, the electronic kill as well as, the laser kill all configured on the same weapon configuration.

A Suggested Way Ahead For Us

With such high proliferation of UAS in our immediate neighbourhood, building and refining capabilities to kill them in the combat domain is an operational need whose time has long come. Some thoughts on proceeding forward in our scenario are offered:

- Our basic premise must remain anchored on achieving cost-effective kill on UAS. Bringing them down using close-in-weapons like air defence guns, man-portable missiles or more 'inappropriately' using VSHORADS (Iglas, Strela, OSA AK SAMs) or SRSAMs (Kvadrat/ Akash) or worse by scrambling air defence aircraft in search-n-shoot missions should remain the last option.
- Going with the flow of realised technology at the cutting edge today, we need to go in for two types of cheap kill solutions, namely, 'hit-to-kill' and laser kill.
- Developing a cheap 'hit-to-kill' system of the type of Peregrine Eagle or Cougar is well within the capability of the DRDO with technological and software support from private industry where many a start-ups are already coming of age.
- We already possess the capability of having a basic pilotless target aircraft (BPTA); the missing component is to design a kill vehicle, either strap-on or as a detachable integral that can be put on a guidance course to the UAS. This guidance package must make use of existing sensors keeping the kill vehicle as cheap and simple as possible (no RF/IR seeker to be configured on board).
- The second and parallel direction in which to move simultaneously is to develop the anti-UAS laser kill system. In this field also, excellent work is already at hand (details not covered). Thismust be taken further with due support from private industry where several players,

both indigenous, as well as, foreign original equipment manufacturers (OEMs) are capable of offering technological capability in the field of lasers. The foreign OEMs could be taken on the joint venture/memorandum of understanding (JV/MoU) route.

- The user Directorate could provide an integrating platform where the already spelt-out felt need can be addressed by public and private partnership.
- It is relevant to bring out that the cutting edge technology today is already addressing the following erstwhile challenges in the domain of laser kill weapons:
  - Capability to aggregate the power of a number of laser sources at the point of impact so as to generate a kill beam of sufficient intensity.
  - Capability to auto-focus multiple beams from co-located sources at a single chosen point of impact (target) in near real time through an 'auto cued-on' arrangement.
  - Use of optics (gimballed mirrors/reflectors and more) on board aerial kill vehicles to reflect multiple laser beams aggregated from different sources on ground as a strong 'cumulated kill beam' (unlimited magazine) on the target UAS.
- Research in the field of charged particle beam and high power microwave must be kept alive by the DRDO as well as a few other players in the private industry working in this vertical so as to come out in the foreseeable future with possible soft kill solutions in this still evolving field.
- For any new procurement of guns and missile systems, anti-UAS capability (especially the sensor capability to detect minimalistic targets) must be checked out before finalising the procurement.
- Old vintage Russian radars operating in lower frequency bands (P-12, P-18, P-19, PRV-16) must not be junked. These are excellent radars that can detect slow moving UAS. Even the sensors on the current holding of GBADWS have shown varying capability of detecting small RCS UAS. These sensors must be used optimally in building cheap kill capabilities (say UAS hit-to-kill system).

- And, lastly, why kill the UAS when the following can be easily done:
  - Make the UAS stream the audio/video data not only to its GCS but also to the hacker station.
  - Make the UAS send deliberately wrong data to its operator.
  - Confuse the UAS with false GPS information, waylay it or direct it in the wrong direction. Just hijack it.
  - Take command of the UAV and make it fly to your wish/make it crash/use it like a guided missile. (remote hijacking).
  - Extract valuable seed data from the built-in threat library of the UAS and use it to your advantage.
  - All these and more are options to be exploited.....

PART 3: MUM-T—THE LATEST BUZZWORD

As stated earlier, Manned and Unmanned Teaming or MUM-T for short, basically relates to that win-win situation in which the human genius of a combat pilot, his tolerance for ambiguity, hunch, intuition, experience, instant decision making capability and more, is amalgamated with the brute lethality, long endurance, needle-like precision and all-terrain and all-weather capability of the modern day UAS in joint missions. This concept is fast catching up across the leading nations of the world since the initial results show multiplicative and disproportionate gains, albeit not without many a challenge; some operational, some ethical.

Going back on the timeline shows that not exactly MUM-T, but something remotely close to it, happened way back in World War II when a B-17 Flying Fortress of the Royal Air Force (RAF) started as a manned mission to bomb a submarine bunker in Germany. After achieving a cruising altitude of 20,000 ft and arming the explosives the two-pilot crew parachuted into the English Channel letting the aircraft fly to its intended target as a robot. Though the mission failed but with it, a step-cousin of MUM-T passed into evolution that would tickle the minds of inventors in the years to come.[32]

As the years rolled by, the power of the unmanned continued to prove its unmatched utility as the 'Dull, Dirty and Dangerous' machines, credited with many an operational virtues, proliferated more and more in the opetraional domain Soon the debate on manned versus unmanned became irrelevant and subject matter experts started to look at the 'AND' solutions instead of 'OR' solutions.

The earliest reports came in of the joint exercises between AH64 D/E Guardian being paired with UAS like the MQ1 Grey Eagle (a derivative of MQ1 Predator) and RQ7B Shadow UAS. Similar exercises were reported using BEL OH 58 Kiowa helicopter being paired in joint exercises with several different UAS.[33]

Manned and Unmanned Teaming

Cutting straight across to 2015, there is an account by an attack helicopter (AH) battalion commander stating that in the Afghanistan war, some 60 per cent of the direct fire missions were helped out by MQ1C Grey Eagle drones enabling Apache crews to see full-motion videos of the areas they were flying into or the enemies they were sent to attack, well before they reached their destinations.[34]

Today's MUM-T enablement not only allows the pilot in the manned platform to view the videos generated by the unmanned machine, but also enables him to control its sensors and its flight path and permits him to fire the weapons carried on board the UAS.[35] This is a phenomenal enablement. Not only restricted to helicopter-UAS teaming, things are moving ahead in trying to achieve MUM-T using combat aircraft and UAS. Reports of the F35 joint strike fighter being capable to control the sensors and payloads of nearby UAS or even controlling a swarm of UAS are doing the rounds.[36]

While the above MUM-T examples are generally related to USA, open sources tell us that MUM-T as a concept, is fast taking shape in other nations leading in aerospace power and dominance. Some examples:

- According to a report, while Russia wraps up its development on the Sukhoi T50 PAK FA fifth generation stealth fighter, it is exploring concepts in the MUM-T domain. Reportedly, the PAK FA will have modifications so as to make it MUM-T capable. Also, reports are already pouring in of future generation Russian aircraft designs (sixth generation and beyond likely to be realisable in the timeframe of 2030-2035) which will all be MUM-T capable. Where is the doubt that MUM-T will see many a grade of future upgradation in the machines yet to come?[37]
- In 2014, M/s Finmeccanica completed a capability concept demonstration for a rotary wing UAS for the UK MoD exhibiting launch, recovery, mission management and mission system integration with manned platforms.[38]
- China is also gearing up in the MUM-T domain in a big way, though specific information about them is scanty on the open source.

Connected intrinsically with MUM-T is the concept of LOI or the level of Interoperability between manned and unmanned platforms. Classified through grades 1 to V, the LOI tells us just how much the manned and unmanned platforms are interoperating with each other and one another. Starting with the bottom line of LOI 1 where only the ground control station (GCS) can receive the UAS data indirectly (implying time delayed verbal reports), the interoperating levels are scaled up in steps. LOI II refers to a real time payload feed. LOI III relates to the actual control of the payload (sensors/weapons) on board the UAS. In LOI Level IV, the manned platform is in total control of the UAS except its launch and recovery while at LOI V even that function is in the hands of the manned platform. The current level of MUM-T operations is hovering between LOI IV and V.[39]

MUM-T as a viable concept is fast acquiring new dimensions as many new things are being tried and newer windows of opportunities opened through comprehensive technological enablement. Some examples:

- When manned and unmanned systems are threaded into joint missions, the cockpit workload of the pilot increases manifold since the man-in-the-loop is now to control multiple machines, some directly and others indirectly. Technology is at play to reduce this workload saturation by letting the UAS do things without pilot interface by building suitable artificial intelligence (AI) suits. These could include maintaining a threshold surveillance level activity as default and auto-reporting variations in levels, picking up targets as forward scouts based on threat library data, fully autonomous flying to destinations, keeping stations based on pre-fed data, auto battle damage assessment and more. The idea is that the pilot manages the multiple UAS payload only by 'exception' where either a human decision interface is required or a manual override is needed in case of forcing a change or averting an emergency.[40] Over time, smart algorithms and AI suites will gradually permit many more battle functions to be taken on by UAS without human intervention. Tasks like mission planning (processing with givens), data collection and analysis are cases in point.

  For better interoperability and to avoid collisions while operating in the same airspace in real time, the best of the line capability is being built into unmanned systems. In 2016, General Dynamics experimented with the idea of fitting a state-of-the-art active electronically scanned array (AESA) radar on board its MQ9 Predator. Such high-end sensors will not only be ideal for aiding collision avoidance but also, will provide capabilities for independent targeting of incoming threats, hunting for potential targets and their selection based on AI suite on board, duly complemented by a dynamically updated threat library. Such systems could also be suitable platforms for ELINT and even act like jammers on demand.[41]
- In line with the thought of UAS to be independent platforms for the end-game, their weaponisation started long back. AIM 9X

Sidewinders, AIM 120 AMRAAMs and AGM 88 ARMs are now mountable on front-end UAS platforms like MQ-9 Predators or M2 Shadow.

- In the early days of MUM-T, the integration between two dissimilar machines (UAS and manned platform) was achieved by engineering an audio visual (AV) communication link that forged a hand-shake of data and video. Technology has taken us a step ahead in which a common data link (CDL) ab-initio provides a common bridge threading a large number of dissimilar platforms for a seamless highway of data sharing. The CDL has built-in security protocols and encryption. The high-end modems provide adequate bandwidth, giving a capability of streaming full motion videos and pictures in real time. The communication and data link that used to be engineered to provide the interoperability to the requisite LOI level is now coming in-built in the MUM-T suite.
- Besides the interface between two aerial platforms, MUM-T has long been happening between the UAS and soldiers and commanders on the ground in the tactical battle area (TBA). As a part of CDL extension, the ground users are getting equipped with such tools that allow them to connect with the UAS for multiple tasks based on authorisations and protocols. It could be simply to request for a motion video relating to terrain of interest or an authorisation to take control of the electro optical/infrared (EO/IR) sensors on board the UAS thus getting the capability to build situational awareness in a directed mode.

Far ahead of just being in network contact, the MUM-T task spectrum is steadily widening as new avenues in mutual enablement are being thought of and practised in joint exercises. A snapshot...

- In joint target engagements, manned and unmanned machines are sharing target accessibility co-ordinates through imagery, sensor data, situational map overlays and real-time spot reports. Complementary weapon and munitions selection ensures maximum multiplicative effect at the target end and auto deconflicted munitions trajectories from multiple airframes ensure their safety

- Auto sharing of post-strike imagery in real time permits decision of repeat strike.
- Enhanced observation and situational awareness envelope is being ensured through gridding the multiple sensors on board fused for multi-sensor tracking (MST) of the same target so as to avoid duplication in reporting of the same target by multiple sensors.

When the MUM-T concept was just about evolving, the 'teaming' per se was essentially between the pilot on board the combat aircraft and the UAS controller at the GCS. Technology-driven, there is a gradual shift taking place in this arrangement in that as the UAS are steadily achieving more and more autonomy in launch/recovery, navigation, observation and sensor/weapon/EW operation they are getting more and more disconnected from the GCS. The connect is now happening increasingly between the combat pilot and the 'resources on board the UAS' (the human-machine interface) gradually working to a state where the UAS will be totally independent of the GCS and will act as a force-multiplier tool in the hands of the combat pilot.

While all the advantages of MUM-T stand on one side, on the other stands the solitary question—'Is the world yet ready for Machine over Man?' measured on the scale of the ever-increasing autonomy for the unmanned platform. The questions that stare us today actually lie in the ethical domain. Some of these run on the following lines:

- Can the unmanned machine be given the autonomy to strike on a target it has identified, based on its AI suite and threat library?
- In an end-game sequence, can a UAS be given self-authorisation to re-strike in case its sensors give a miss/inadequate kill feedback?
- Can a UAS be allowed to leave the formation, albeit for a brief period, in order to take on an identified target of opportunity? And, above all...
- Can an unmanned machine lead a MUM-T mission? (this lead is not implying a scout role but a proper command lead)

The present-day answer to all these and more is an emphatic NO as the manned community is firmly fixed on 'seek permission before every strike' and the invariable override function of 'Man-in-the-Loop' across board. Well, this is the position as of today, but who has seen tomorrow? Concepts and mindsets may change, who knows? Time has the key!

MUM-T in India

Taking our own baby steps, we have indeed come a long way in our nearly two-decade old journey when the Indian Army in 1997-98 acquired the Searcher Mark 1 UAS from IAI Malat to be quickly followed by the other two Services. Over time, came the Searcher Mark II and the Heron UAS. However, all these have been dedicated to very basic and traditional UAS roles encompassing surveillance, reconnaissance and intelligence gathering missions from the northern and eastern high altitude areas to the southern deserts (DRDO UAS products not covered). Cut to the current era, and examine the following open source reports:

- The Defence Acquisition Council in its meeting on 25 June 2016 has cleared the acquisition of 40 Predator surveillance drones from the US company, M/s General Atomics. These drones are meant for the Indian Navy for surveillance tasks in the Indian Ocean Region (IOR).[42, 43]
- The *Indian Express* reported on 8 April 2016 that the IAF has also asked Washington about acquiring 100 Predator C UAS to carry out strikes on militant camps.[44]
- The *Economic Times* reported on 28 September 2015 that India and the USA have signed a USD 3 bn deal for the purchase of 22 AH 64 Apache attack helicopters and 15 Chinook heavy-lift helicopters from M/s Boeing.[45] This is Apache Block III configuration; stealthy, versatile, designed for all kinds of missions with laser and infra-red systems for day-night operability and armed adequately with Hellfire missiles, 70mm rockets and an automatic cannon.

What does all the above indicate? Surely and steadily we are going into platforms that can support the emerging concepts like the MUM-T. But MUM-T does not happen overnight. Concepts have to be evolved, network-centric communications on the lines of CDL have to be developed step-by-step that bridge the divide between the UAS and the AH seamlessly, critical skill-sets have to be identified and crews have to be trained both in the manned and the unmanned domain and specific battle drills have to be perfected over years and years of training. And what position do we have on the ethical baggage as mentioned above? That indeed is a long haul.

Be that as it may, the author is of the view that it is high time to commence building the conceptual framework for MUM-T operations. Think tanks like the Centre for Joint Warfare/Land Warfare/Air Warfare Studies (CENJOWS/CLAWS/CAPS) could start putting in place the thought process duly supported by the three Services, DPSUs, DRDO and the industry. Services training schedules could subsequently get re-structured over time to accommodate new concepts and doctrines. Once baby steps are taken, the pace will build up in the years to come.

Going by the cliché—The journey of a thousand miles begins with a single step.

NOTES

1 Aviation History as Global Hawk Completes US-Australia Flight: Australian MoD Press Release in http//www.en.wikipedia.org/wiki/Northrop Gruman RQ-4-Global Hawk. Accessed on 02 Jan 2018.

2 http : // www.reuters.com /02/ Pakistan-USA-drone>taliban-militants-say-they-shot-down-US-drone. Accessed on 02 Jan 2018.

3 http://www.foxnews.com/drones-vulnerable-to-terrorist-hijacking-researches-say. Accessed on 02 Jan 2018.

4 http://www.navy.mil/submit/display.asp?story_id66319, navy-closer-to-landing-UAS-on-aircraft carrier. Accessed on 02 Jan 2018.

5 http://www.news.US on.org/news-analysis/news/unmanned-joint-strike-fighter? Accessed on 02 Jan 2018.

6 Autonomous Fire Scout UAS Lands on Ship,http://www.aviation.week.com/article.aspx?id = xml/asd-08/requirement-for-larger-fire scout-VTOL UAS. Accessed on 02 Jan 2018.

7 Weisshaar T.A (2006) morphing-aircraft-technology-new-shapes-for-aircrafts-design, http://www.rto.nato.int/abstrach.asp

8 http://www.en.wikipedia.org/wiki/orinthropter. Accessed on 02 Jan 2018.

9 http://www.ho.secs.vcla.edn/np-contract/uploads/04/Jpl10-2001.pdf. Accessed on 02 Jan 2018.

10 http://www.usavision.com/2011/11/21. Accessed on 02 Jan 2018.

11 http://www.army-technology.conc/projects/rq11-raven. Accessed on 04 Jan 2018.

12 en.wikipedia.org/wiki/pseudolite. Accessed on 04 Jan 2018.

13 http://www.designers.com/another.asp?section – 'autonomous-UAS-fly-in-swarm'. Accessed on 04 Jan 2018.

14 http://www.blogs.bn.edn/bioaerial 2012/2012/.../swarming-a-team-sport-for-UASs. Accessed on 04 Jan 2018.

15 http://www.sciencealert.com>first-ever-drone-swarm-attack-has-struck-russaim -military-bases. Accessed on 04 Jan 2018.

16 https://www.cnbc.com>swarm-ofarmed-DIY-dronesattack-russian-military-base-in-syria. Accessed on 04 Jan 2018.

17 Ceejun Chandran. Defence Pro Ac Biz News, Vol 5, Issue 5, Sept-Oct 2017>unmanned-aerial-vehicle

18 ibid.

19 http://www.economictimes.indiatimes.com>drdo-successfully-test-fires-rustom II. Accessed on 13 May 2018.

20 http://www.detect-inc.com/security/html/applied-radar-technologies. Accessed on 04 Jan 2018.

21 http://www.zmnl.her/tanzekek/etic/konferencia/april 2001/prokob.html/surveillance of - UAS. Accessed on 04 Jan 2018.

22 Ftb.rta.nato.int/public/public fulltext/RTO/MP/RTO/esm-sensor-for-technical-information-in-air-defence systems

23 http://www.ieeexplore.ieee.or/ie!5/4803720/4806498/04806528.pdf/low-cost-acoustic-array-function-small-UAS-detection and tracking. Accessed on 11 Jan 2018.

24 http://www.internet data.info/.../UAS-decoy-strategies-theories-and-the-modern-art-of-war. Accessed on 11 Jan 2018.

25 http:/www.popsei.com/when-drones-go-wild-airfoce-shoots-them-down. Accessed on 11 Jan 2018.

26 http:/www.state.com/Israel-shoot-down-drone-that.may-have-been-made-by-Iran. Accessed on 11 Jan 2018.

27 http://www.guadian.co.UK>Glen Green World on security and liberty. Accessed on 11 Jan 2018.

28 Dave Mazumdar, "Iran Captured RQ 170: How Bad is the Damage?" http://www.en.wikipedia.org/wiki

29 http:/www.howtokillUAS. Accessed on 11 Jan 2018.

30 http:/www.news.c net.com>news>cutting edge>boeing-trucks-ahead-with-8-wheeling-laser-weapon. Accessed on 11 Jan 2018.

31 http:/www.raytheon.com/newsroom/feature/stellant groups /laser-area-defence-system. Accessed on 11 Jan 2018.

32 http://www.armyaviationmagazine.com >archive 5. Accessed on 11 Jan 2018.

33 http://www.wp.nlr/.../onbemand-vliegtuig-rpasuas-veilig. Accessed on 11 Jan 2018.

34 http://ww.flightglobal.com. Accessed on 11 Jan 2018.

35 http://www.breaking defense.com>2015/01.mum-t Accessed on 11 Jan 2018.

36 http://www.tge diplomat.com.2015/05.will-f35-pilotscommand-a-swarm of-drones-in-flight Accessed on 11 Jan 2018.

37 http://www.nationalinterest.org>blog.the buzz> revealed -russia's-lethal-seventh -genetaion fighter jets. Accessed on 13 Jan 2018.

38 http://www.plz.swidnik.pl>uk-mod-rwuas-capability. Accessed on 13 Jan 2018.

39 http://www.cdlsystems.com>VCSlevels-of-interoperability. Accessed on 13 Jan 2018.

40 http://www.mil-embedded.com.mum-t.us-army-uas-roadmap-2016-35. Accessed on 13 Jan 2018.

41 http://IHS Janes Defence Weekly, Volume 50, Issue 34.

42 South Asia Defence Strategic Review, Volume 10, Issue 3, Jul-Aug 2016.

43 Defence ProAc Biz News, Volume IV, Issue 3, May-Jun 2016.

44 http://www indianexpress.com.india-in-talks-to-by-us-predator-drones. Accessed on 13 Jan 2018

45 http://wwweconomictimes.indiatimes.com. Accessed on 13 Jan 2018.

CHAPTER 6

Ballistic Missile Defence (BMD): The Indian Scene

Dr V K Saraswat, Member Niti Aayog

This chapter flags the operational need for a ballistic missile defence (BMD) system for India. At the onset, it establishes the felt need for such a system in the face of the ballistic missile threat that we face today from our potential adversaries. Against this backdrop, it unfolds the status of the indigenous BMD system over the years and examines the linkages it may have with the planned procurement of S 400 system in the future. In the end, a way ahead is suggested for the early operationalisation of the BMD system in the foreseeable future.

Establishing the Felt Need

Let me begin by questioning the premise; why a BMD system for India? The short answer is that in a typical 'cause-effect relationship' in our immediate and extended neighbourhood, the missile threat we face makes it imperative for us to have the capability to defend our national interests and strategic vulnerabilities against inimical forces. While it goes without saying that all nation-states in our immediate or extended neighbourhood are not potential threat states, the question here is to match capability-vs-capability, commensurate with our perceived world status in the foreseeable future, besides countering the missile threat from our potential adversaries.

While on the missile threat, it is stated that the capability of a nation state to hold out a viable threat is not defined by the mere possession of ballistic missiles (BMs); the pillars of strategic capability include a trio of 'nuclear warheads', 'delivery means on land sea and air,' and 'command, control aspects including field infrastructure'.

Seen in the above light Pakistan, with an arsenal of BMs from 80 to 3,000 km[1] has effectively set up an equation of deterrence against our superior conventional and nuclear capability duly augmented with its friendly nexuses (North Korea, China, Libya, Iran, Saudi Arabia, etc). Most of the BMs are capable of carrying strategic warheads.[2] As to the trio mentioned above, the associated command and control structure by way of a National Command Authority, Strategic Forces Command, Missile Brigades and enabled delivery means are also reportedly in place. The details of these have already been covered in Chapter 1 (ed.).

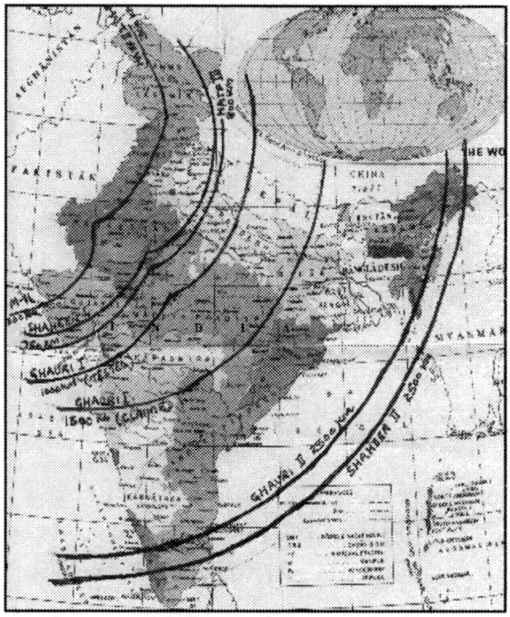

On the other hand, China has an awesome strategic capability with all the three legs of the triad already in place. With ranges of its BMs extending right up to Australia, the reach into and hence the vulnerability of, the Indian subcontinent is complete. The associated command and control structures including such force multipliers as strategic and space-based surveillance capability, battle management systems and a large array of warheads, give total credence to the strategic deterrent held by China.[3, 4] The latest type of threat developing in the context of ballistic missiles from China are the hypersonic missiles (e.g. DF 26), the counter for which must also operate in the hypersonic dimension. The details of such a threat have been covered in Chapter 1 (ed.).

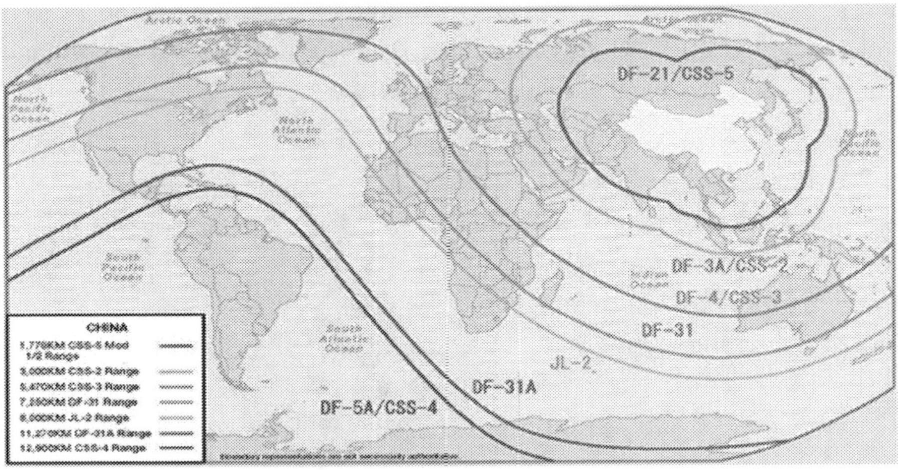

Chinese SSM Capabilities

In the face of the above stated 'Threats-in-being', our policy of 'No First Use' (NFU) itself mandates that we must be capable to survive the first strike by taking on the adversary's BMs in order to be able to respond with massive punitive retaliation' causing "unacceptable damage", as stated in our nuclear doctrine This, translated in threat terms, implies that we need to protect our strategic vulnerabilities like select city centres, seats of power/government, strategic command and control set-ups besides industrial, military and nuclear hubs, from the adversary's likely first strike. Besides being "threat driven" alone, we in consonance with the stature of our comprehensive national power also, need to belong to that elite group of nations that possess the capability of ballistic missile defence. So much for the rationale.

The Case for Indian BMD System

A realisation of the above strategic need was firmed up in the Indian security establishment some time in 2000-2001. A world scan revealed that all the major powers of the world, viz., the USA, Russia, China, etc., had hierarchical BMD systems covering diverse range/height bands: Russian

ANTEY 2500 (75 km) to Almaz Antey S-300/PMU (100-400 km), US Patriot from 87 km to THAAD (200 km) and beyond to AEGIS BMD systems. Keeping in mind the analysed threat in our scenario, experts opined that the own BMD system will also need to be hierarchical, covering the range and height band corresponding to the perceived SSM threat to our vulnerabilities.[5, 6]

S-300 PMU-1

Patriot

Thaad

Aegis

Keeping in mind the indigenous technical prowess of the DRDO, the question of indigenous system versus eximport was decided in favour of the former. It was also decided that our system will have to be effective both in the endo, as well as, the exo-atmospheric regions making a beginning with 'Terminal Phase Interception.' Thus was born the Indian BMD programme spearheaded by the DRDO.

This programme (code name: Programme AD) was to be developed in two phases. Phase I was to build the BMD shield against an incoming threat of ranges up to 2,000 km, while Phase II extended this capability to cover the missile threat from 2,000-5,000 km. The timelines for completion of Phase I was 2012 and for Phase II by 2016.

The strategy for interception was to follow a typical air defence engagement sequence. In that, the incoming missile was to be detected by long range tracking radars (LRTR) of a range of 500-1,000 km. Once classified as hostile and prioritized for engagement, the target information based on its heading parameters and the predicted point of impact was designed to be passed to a multi-functional radar (MFR). The MFR tracking the incoming threat was to continuously pass the above target parameters to the missile via a secure target up-link transmitter (TUT) to be received by the missile through its target uplink receiver (TUR). At the opportune moment, the interceptor was to lift from the launcher and be guided to the target on the TUT-TUR link. Once in the end-game, the target was to be picked up by the active seeker on board the missile which was to take the missile to the target for a catastrophic kill.

All the elements were designed to be interconnected through redundant communication links. Since the focus in Phase I and Phase II was both endo and exo-regions, interceptors were designed accordingly. For Phase 1, Prithvi Air Defence (PAD) was to be the interceptor for the exo-atmospheric interception up to altitudes of 50-80 km while advanced air defence (AAD) was meant for interception in the endo-atmospheric region up to an altitude of 30 km. PAD is a two-stage liquid fuelled interceptor. Its future version is Prithvi Development Vehicle (PDV) in which the liquid fuel propulsion stage is replaced by a solid fuel stage.

Phase II was to have two interceptors, AD-1 and AD-2. Roughly, the Phase II systems were supposed to match the capability of the THAAD missiles deployed by the USA. Such a missile can intercept targets up to 200 km, altitude and track targets at ranges in excess of 1,000 km.

As per open source data, the entire system consisting of weapon and support systems configured in Phase I and Phase II stands fully configured and tested. As of August 2017, the Government is reported to have cleared the allocation of land for setting up of the radars and support systems of Programme AD.

Programme AD is likely to be deployed for the protection of strategic and prestigious assets such as city-centres/seats of power/military, industrial and financial hubs, economic/strategic targets like big oil refineries, national oil storages, sites of nuclear weapons, national command posts and more. The huge swaths of the area air defence and anti-missile coverage which will be available by one system deployment will cover several of these vulnerable areas at the same time.

Programme AD will also find a connect with the S 400 system for which clearance has been accorded for the procurement of five regiments from Russia. S 400 is also an anti-missile system and will therefore complement the indigenous capability as contained in Programme AD. In that, the fire units (FUs) of Programme AD system and S 400 system could be deployed complementing each other.

It is however to be appreciated that given a huge quantum of VAs/VPs that need to be protected, the limited quantities of Programme AD FUs combined with FUs of S 400 will still leave voids. In any case, S 400 is still several years away into the future as its FUs start to get inducted and operationalised in a phased manner.

Imperatives

Keeping in mind the net ballistic missile threat in our neighbourhood that exists as of now, the case for BMD for India cannot be over-emphasised. The need of the hour is to move forward expeditiously towards realisation of Programme AD systems as deployable FUs under Strategic Forces Command (SFC).

In that, as the Indian BMD system moves towards operationalisation, certain associated points need serious consideration.

First and foremost, the declared NFU doctrine is essentially based on the premise that we have the capability to take on the enemy's first strike and survive to cause the stated massive punitive retaliation. This would call for building robust and survivable infrastructures, fail-proof and survivable regime of codes and authorisations for weapon delivery, duly supported by several layers of redundant and secure communications. Besides the wherewithal to stage a response, the survivability of actors also needs to be ensured through alternative command posts with a chain of hierarchy fully defined and known to the decision makers up and down the chain.

As for collateral damage and casualties to the civil population, the enemy's first strike is likely to be a calamity, very much like a man-made disaster as explained by the guidelines issued by the Natural Disaster Management Authority (NDMA). These comprehensive and very well documented guidelines which call for a whole lot of actions at the Centre and State levels, need to be "realised on ground" and practiced and rehearsed time and again till the e potential victim population becomes fully aware of the looming danger.

Planned dual use of existing infrastructure, like construction of mass shelters, keeping disaster relief bricks at hand, catering for disaster medication and many other calamity management measures as defined and laid down in the NDMA guidelines call for immediate attention and action by the concerned authorities and organisations.

Last, but not the least, will be the requirement to put into place an integrated command control and battle management system which will integrate multifarious 'sensors' and 'shooters' of air defence and anti-missile system into one seamless loop. This life-line system will be responsible for optimising surveillance resources through multi-sensor fusion in the detection of an incoming threat followed by target designation and weapon selection so as to ensure successive punishment to the incoming threat. The basic tenets of these systems are steadily being put into place. Programme AD and later S 400 systems will have to integrate with the same.

As to the planned operational deployment of Programme AD, open sources reported way back in 2012-13 that once the programme is operationalised (target-2016, likely by 2020-21) the system FUs will be deployed to provide BMD cover to the two metro cities of Delhi and Mumbai. Since the system operationalisation for Phase II was nearly getting complete in 2017, it was reported in *The Times of India* in March 2017 that plans are afoot for deployment of BMD FUs for New Delhi.

The requirement in the immediate will be to ensure that the operationalisation of the system moves forward which essentially implies the following:

1. The system must completely and exclusively change hands 'from the DRDO scientists' to the 'users and stakeholders.' In that, the Strategic Forces Command must fully own, operate and actualise the system as a weapon in its arsenal. Full operationalisation must be achieved at the earliest.
2. The sequence of deployment of Programme AD FUs, which involves a large number of actions demanding multi-disciplinary and multi-agency coordination must start to unfold immediately.
3. Sequence of action to realise associated command and control infrastructure, missile testing and storage sites and tuning up the disaster management response must come up alongside

NOTES

1 https://www.ewn.wikipedia.org> list of missiles Pakistan. Accessed on 14 Jan 2018.

2 https://www.thediplomat.com>2017>01> pakistan missile research and development programme. Accessed on 14 Jan 2018.

3 https://www. globalsecurity.org>missile> chinese missiles. Accessed on 14 Jan 2018.

4 https://www.surface-to-air-missiles-of -peoples-republic-of-china. Accessed on 14 Jan 2018.

5 https://www.fightersweep.com>hoe-scary-is-s400. Accessed on 14 Jan 2018.

6 https://www.the diplomat.com>what-is-thaad-what-does-it-do—why-is-china-mad-about it. Accessed on 14 Jan 2018.

CHAPTER 7

Dealing with the Stealth Threat

Brig GS Choudhry, VSM

The stealth muscle in the air threat vehicles of today is one dimension of survivability that is fast becoming a nightmare for air defence warriors. Why so? Simply because survivability directly translates into lethality. The relationship is fairly simple. Stealthy machines, by managing to become near- invisible to the sensors of the adversary, not only survive by avoiding the onslaught of the defender's air defence arsenal but also, with all the smart and intelligent arsenal at their disposal, cause disproportionate damage by getting to attack the defender's vulnerabilities nearly unchallenged (read undetected). Shadow boxing of sorts! Detection of stealthy air threat vehicles is therefore an operational compulsion.

The Stealth Threat We Face

Has the stealth threat arrived at our doors? Yes, for a one-word answer, if we were to examine some features of the frontline aircraft with our northern neighbour that are fast approaching operationalisation and are making waves in professional circles.

The case in point is the Chengdu J 20 (Black Eagle), a stealthy fifth-generation fighter aircraft (FGFA) with the People's Liberation Army Air Force (PLAAF) optimised for long range strikes. Starting its first flight on 11 January 2011, the J 20 was planned for induction in January 2018. In March 2017, the Chinese media reported that the aircraft had entered initial operational capability phase with the air force.[1, 2] In September 2017 it was reported that the J-20 had officially entered military service with the PLAAF.[3] In October 2017, the Chinese media reported that Chengdu Airspace Corporation (CAC) had initiated series production for the J 20 and was on the path towards achieving full operational capability.[4] Though very few details are available about the said FGFA, the stated (and unstated/analysed) stealth muscle of the J 20 may run like this:

J 20 Stealth Fighter

- A careful inspection of the aircraft structure shows the leading edges of the wing and the end tail surfaces at the same angle. The same is the case with other small structures, viz., the air intakes, bypass doors and air-refuelling apertures. This design feature is called planform alignment (also in F 22 Raptor). This sort of alignment returns the defender's radar signals in a specific direction away from the radar emitter (thus avoiding detection) rather than returning a diffused signal which can be detected by several anti-stealth radars (bi-static, multi-static).
- Though the latest prototype rolled out wearing only the yellow factory fuselage primer, the Black Eagle/Soaring Dragon will in all probability get a state-of-the-art radiation absorbing material (RAM) coat in order to send the minimalistic radar cross-section (RCS) signature. One of the earlier pictures of the aircraft showed the Haze paint RAM which was used on F 16s. Open sources reported

in November 2015 that the Chinese Huazhong University of Science had developed a RAM wherein a metal slab is placed over a layer of thin metal honeycomb under which there were several layers of ultra-thin active frequency selective surface materials, each optimised to absorb a range of radar frequencies. Will such a thing find its way on board J 20? While the answer could be a 'yes' or a 'no,' it is certain that the RAM on board the J 20 will have cutting edge features like the use of nano-materials, active carbon coats or even a plasma-driven stealth coat.

- Another thing to note in the J20 are the very sharp edges, vertical cuts and an equally sharply pointing nose with a frameless canopy. These are deliberate design features which ensure that the incident radar energy of the defender's radars is never returned in a conventional mode but is invariably deflected at an angle away from the mother radar, making detection difficult. A You Tube video of this aircraft in flight throws up an interesting observation. It is seen that throughout its flight (except for a brief moment when the aircraft is overhead and offers a fleeting view of its underbelly), the shape of the aircraft is such that it never offers a flat perfectly reflecting surface in manoeuvre. The axi-asymmetric nozzles with jagged edges are low observables which effectively conceal the engine exhaust; the main aiming point of infra red (IR)-seeking SAMs. These features not only make the aircraft a low observable to defender's radars but also poses a very difficult target for the heat-seeking surface-to-air missiles (SAMs) fired at it in the visual/beyond visual domain.
- Supersonic combat jets have another issue that relates to gas turbine engines (GTEs). Such engines require a constant subsonic air inlet flow since these are unable to handle supersonic air speeds. This is so because the shock waves created at such speeds may result in dangerous vibrations in the turbine blades which can not only, result in reduced thrust but also, may lead to a catastrophic engine failure. The air required for the GTEs which is resident at the aircraft boundary (boundary layer) has therefore to be reduced from

supersonic to subsonic speeds. This is done by using diverter (splitter) plates. These attachments sitting next to the aircraft skin serve as perfect reflecting bodies for the incident radar energy giving away the location of the aircraft. Technology has now got rid of the diverter plates in offering diverter-less air inlets (DSI). The J 20 has a DSI where a bump and a forward swept inlet cowl deflects the boundary layer (supersonic) air flow away from the aircraft's engine. This makes the aircraft a real low-observable machine.

- There used to be a time when the external strong points under the aircraft fuselage carrying a known inventory of armaments were a sure give-away of the aircraft's identity. As stealth caught on, the aircraft pulled in their weapon loads into their internal bays. However, during the precise moments of firing (also the same moments when the defender's sensors and weapons were ticking), the weapon bays used to open to permit firing. This also opened the entire arsenal load at a critical time. Then came the canted trapeze.
- When it was time for firing, the canted trapeze would lower one missile into the air stream at a time without directly exposing the entire bay. If the missile has to achieve a lock-on-before-launch (LOBL), it will be lowered till lock-on is achieved. For lock-on-after-launch (LOAL) missiles, the lowering duration will be shorter (F 22 Raptor). In the earlier designs, once the canted trapeze was lowered and till the time the firing took place, the entire weapon bay was perforce visible to a potential attacker who could have a view of the entire magazine. The J 20, does one better in this design. Once lowered, it permits the weapon bay to be closed even before the firing is to take place thus permitting only one missile exposure at a time without a view of the whole magazine for extended duration till the firing event. This adds to the stealth muscle of the aircraft.
- Another smart idea is to have two types of weapon bays. One main weapon bay storing one lot of short and long range AAMs [PL9 (22 km), PL12 (70 km) C/D and PL 21 (>100 km)] and another two smaller lateral weapon bays (smartly tucked behind the air

inlets in the low visibility domain) containing only the short range P 9 AAMs. The logic of a smaller bay yielding smaller exposure and permitting quicker reaction adds to the stealth muscle of the machine.

Subject matter experts (SMEs) are comparing the stealthy muscle of the J 20 with the likes of F 22 Raptor, F 35 Fighting Falcon, Su-PAK FA the Japanese ATX or Korean KAI KF-X (the last two still under development). The question is not which aircraft tips the other or otherwise; the grave fact is that the stealth threat in a tangible measure has arrived at our door. An expert analysis has opined that the J 20 has the potential of a high performance stealth aircraft. With the ever strengthening military bond between our neighbours in the north and the west, the likelihood of seeing these machines in our west is only a matter of time.

The Shenyang J-31 Gyrfalcon or Falcon Hawk is also maturing alongside as a fifth-generation twin engine, mid-size, stealthy, multi-purpose medium fighter. Starting with its first flight in October 2012, the aircraft is due for induction in 2018-19. It is likely to have comparable stealthy features as quoted above; DSI bumps, stealthy weapon bays and mat-finish RAM, to name a few.[5]

J 31 Gyrfalcon

If that is the shape of things firming up, where is the doubt that the stealth threat has indeed arrived at our door. As stated, the same, in the years to come, is also likely to make its unwelcome appearance on our western border.

Dealing with the Stealth Threat

If we have to counter the deadly threat from such multi-role stealth fighters like the J 20, J 31 and others, we need to have sensors that can detect them in the first place. Thanks to many a state-of-the-art stealth technologies on board (Planform Alignment, latest RAM, stealthy exterior engineering, stealthy weapon bays and more), stealth aircraft remain undetectable by the old conventional radars. Also, since most of such class of aircraft have long standoff capabilities, they are likely to strike while remaining in the BVR (Beyond Visual Range) domain. It follows that if these machines have to be countered, they must be detected electronically by our sensors at long ranges. How do we do it?

Basically, when the electromagnetic (EM) waves from the surveillance and other radars of the defender fall on the stealthy (also called low observable) aircraft, they either get absorbed or diluted or deflected in one or multiple directions away from the mother radar. This results in the mother radar getting very weak to nil reflected signals from the aircraft, thus remaining unable to detect the stealth aircraft and paint it on its screen.

Air defence warriors the world over are working to break the above trick one way or the other (going by the name of counter very low-observable or CVLO technologies). This is very briefly the world snapshot of the CVLO scene:

- The Russians are in the forefront, followed by the Americans, Israelis and the French who are working on radar frequency bands that are optimised to detect stealthy targets. Experts have realised that lower frequency bands (VHF and low UHF ranging from 0.2-0.5 GHZ) make much better CVLO radars than the higher frequency band I/J/K or X/Ku/K Band (8-20 GHZ) which are conventionally the most popular radar bands for military application.[6]
- In rudimentary terms, the equation is simple. While the higher band radars are more accurate and more responsive with smaller antennas, their power handling capability is limited. Also, their waves suffer far more atmospheric attenuation and absorption than low band radars. Due to this, high band radars do not make good

stealth radars simply because detection of stealth requires the waves incident on the target to have good signal power with minimum attenuation by the atmosphere. Only such waves will have adequate residual reflective power after a big chunk of it is either absorbed or attenuated by stealth aircraft. VHF and low UHF band radars make a nice CVLO choice because, firstly, they can handle large transmission power and, secondly, the waves exhibit minimal absorption/attenuation by the atmosphere. The price to be paid is in the form of huge and unwieldy antennas and less accuracy in target resolution.

Russian NEBO M 3 D radar

- The Russian 55Z46M Nebo M 3D radar or the Chinese KJ 2000 and KJH 200 are CVLO radars in this category. In fact, these radars go a step further. They feature not one, but multiple bands; Low band (VHF) for anti stealth and high band (L and X bands 40 GHZ+) for other accuracy-driven tasks like illuminating the target or tracking it post detection. They are thus also called MIMO (multiple input multiple output) radars.[7]
- For threats that remain on the other side of the hill and use it as a stealth feature (a capability that exists with our immediate neighbours), VHF/HF radars also make good over the horizon (OTH) radars. They launch their waves at higher elevations (over

the hill) escaping out of the atmosphere and hitting the ionosphere from where they get reflected in hops covering areas beyond the line of sight (LOS) obstacles.

- Experts the world over, are also exploiting the lasers for detection of stealth targets. Laser waves, because of their short wavelength, high beam quality, strong direction-ability and high measuring accuracy, can successfully detect stealth targets. Such waves are ideal in initially identifying the target and positively displaying it, thereafter following it up with orbit recording in a continuous mode. Due to their high coherence and high resolution, laser waves also possess anti-jamming capabilities. Radars using laser waves are also called LIDARS (light detection and ranging).[8]
- Since any stealthy target will either try to absorb the radar energy or reflect it back in a pure/diffused/altered condition away from the mother radar, the trend to counter this trick is to transit from mono-static radars where the transmitter and receiver are at the same place to bi-static/multi-static configurations in which the radar receiver(s)/is/are placed at a different location(s).
- Going a step further, an open source mentions the Chinese going in for an integrated air defence system hooked to a central information processing centre which features multiple radars with high altitude drones (Divine Angle) along with satellite tracking. Based on the theme that a target cannot remain stealthy at all angles, this system claims that stealth targets can 'RIP'.[9]

Chinese Divine Angle Drones

- Another prevalent concept is passive detection or, more precisely, passive coherent location (PCL). In this, the radars do not transmit any radar energy themselves but achieve detection of targets based on receiving EM waves that hit the stealthy targets from multiple domains such as FM, digital audio broadcast or DAB and DAB-T (terrestrial). Lockheed Martin's Silent Sentry ushered in the era of PCL. M/s EADS first displayed their Cassidian PCL radar in the Paris Air Show in 2013. According to a report in *Defense News*, the Chinese are claiming that their DWLoo2 passive detection radar system will make the stealth features of front-runners like the F22 and F35 obsolete.[10, 11]

Chinese DW Loo2 Anti Stealth Radar

- There is yet another niche development called quantum radar which holds the future promise of a strong anti-stealth sensor. It makes use of the quantum properties of photons. Normally a stealth target alters the radar signal and sends a spoofed signal making the defender believe that either the target is harmless (say, sending back a bird signal) or it is elsewhere. The moment spoofing is attempted on photons, these particles undergo a high degree of polarisation. The measure of the strength and extent of polarisation lets out the invisibility of stealth machine.[12]

Focus India

While much is happening all around the world, what does it mean in our scenario? A possible way forward is suggested as under:

- It is fortunate that the latest version surveillance, tactical and missile guidance radars, viz., three-dimensional central acquisition radar (3D CAR), three-dimensional tactical control radar (3D TCR), low level light weight radar (LLWR) and others have significant technological ability to detect stealth targets. Such capabilities are not band-specific alone (high band/low band) but are spread across various other enabling features. (Frequency bands and exact features not quoted).
- Since a large number of ground-based air defence weapon systems (GBADWS), spread across sensors, shooters and battle management systems are at various stages of the procurement cycle, it needs to be ensured that all futuristic systems must invariably have the anti stealth muscle.
- Several legacy tactical control radars (those that can electronically designate target tracks to other radars), early warning radars (those that carry out only surveillance with no capability to designate targets to other sensors), fire control radars (controlling the fire of GBADWS) and missile guidance radars of Army Air Defence (exact names not quoted) are gradually on their way out either because their successors have arrived or their product-improved versions are ready. It is to be ensured that incoming radars have a degree of anti-stealth muscle (which does not necessarily means only lower band radars).
- Certain legacy SAM systems of Russian origin held with Army AD still possess the old vintage P series of radars which can detect stealth targets. Even if the main equipment gets obsolete and is to be de-inducted, these anti-stealth gems need to be kept and exploited as long as it is possible.
- Over a period of time, two DRDO labs, namely, LRDE (Electronics & Radar Development Establishment) and DRDL (Defence Research and Development Lab) along with Bharat Electronics Limited (BEL) have gained considerable clout in the design and production of a whole spectrum of radars. It is time for them to turn the focus on radars with anti-stealth capability. A beginning

can be made by developing a bi-static radar. Future R&D areas could be along the path of PCL and LIDAR. In this, it is not necessary to re-invent the wheel; a head start is available along multiple routes of joint ventures/memorandum of understanding/transfer of technology (JV/MoU/ToT), etc., with foreign original equipment manufacturers (OEMs) under the Make in India initiative.

- In the context of producing anti-stealth equipment, what applies to the defence public sector is also equally applicable to private industry. Since the stealth muscle in our scenario is just about maturing, it is only going to become a more and more prominent feature of the futuristic air threat. Opportunities to develop stealth detecting radars for indigenous consumption, as well as for the export market either on the 'go alone' mode, or through technological tie-ups, is thus an emerging business opportunity.
- In the Def Expo 2016, OIS AT displayed four radars claimed to be developed by them (bird detection, multi-mode UAV detection with a sense-n-avoid function, foliage penetration and portable ground surveillance). There is a need to build further on these capabilities and grow into the anti-stealth domain.

Such is the exciting and dynamic field of stealth and anti-stealth. The eternal cause-effect duel between the prosecutors of air threat and defenders is only going to become more intense and interesting in so far as it pertains to the 'magic of invisibility' at one end and the resolve to 'find the invisible' at the other.

NOTES

1 https://wwwscamp.com/news/china/article/207732/china's-j20-stealth-fighter-flies-fighting-forces-says-stat-media. Accessed on 04 Jan 2018.

2 https://www.en.wikipedia.org. ChengduJ20. Accessed on 04 Jan 2018.

3 Popular Mechanics 29 Sep 17 >china's-J20-stealth-fighter-is-operational. Accessed on 04 Jan 2018.

4 Jane's Weapon Systems 26 Oct 17> cina's-J20 -fighter-moves-into-series-production. Accessed on 04 Jan 2018.

5 https://www. nationalinterest.org>blog>america's-F35-stealth-fighter-vs-china's new j 31 who wins. Accessed on 04 Jan 2018.

6 https://www.ausairpower.nert>APA-CVLObrief>russian-technological-strategy-for-counterVLO sensors. Accessed on 04 Jan 2018.

7 https://www.rt.com>news>russia-deploying-next-generation-nebo-m-radar-complexes to-counter NATO threat. Accessed on 04 Jan 2018 .

8 https://www.metak.de/doppler-lidar. Accessed on 04 Jan 2018. Accessed on 04 Jan 2018.

9 https://www.independent.co.uk>news>how-much-of-a-threat-china's-new-high-flying-drone-to=us-air-superiority. Accessed on 04 Jan 2018.

10 https://www.ainonline.com>aviation-news>EADS-develops-counter-stealth-technology-with spinoff-for-ATC. Accessed on 04 Jan 2018.

11 https://www.ieeexplore.ieee.org>multiband-multiplivative-passive radar-system. Accessed on 04 Jan 2018.

12 https://www.popsci.com>china-says-it-has-quantum-radars. Accessed on 04 Jan 2018.

CHAPTER 8

Challenges of Sustaining a Legacy Inventory

Lt Gen Ram Pratap, (Retd) , PVSM, AVSM, VSM

The most-quoted statistics in the context of ground based air defence weapon systems (GBADWS) of our three Services is the percentage of their obsolescence. The same, till quite recently, was a whopping 97 per cent. This figure is hopefully poised to come down a little bit keeping in mind the recent inductions in the field of sensors, battle management systems, new target systems for live air defence firing, state-of-the-art simulators, training aids and, above all, the commencement of the upgrading of mainstay weapon systems (L70 guns, Schilka weapon system, ZU 23 guns). Various subject matter experts (SMEs), based on the current state of vintage and obsolescence of GBADWS, have opined that the current arsenal will be incapable of countering a contemporary as well as the futuristic air threat. This realisation has led to a consistent effort on the part of the Ministry of Defence (MoD) to revamp the GBADWS of the three Services following a twin-track approach of upgrading the existing arsenal as well as induction of new weapon and support systems.

While up-gradation will happen only gradually and any future induction of GBADWS demands a run time of 7-10 years before operationalisation, the biggest challenge of the day therefore is to sustain the existing (legacy) inventory and keep it operationally relevant, come what may.

Incidentally, the compulsion of sustaining a vintage weapon inventory is not only typical of us (though our percentage of obsolescence is high), countries around the world have been carrying on with their vintage inventory. A snapshot:

- The Russian inventory of GBADWS still carries (has carried) weapons like ZSU-37, ZU-23-2 and ZSU-57-2, besides others. Of these, while ZSU-37, a 1943 vintage self-propelled (SP) gun was discontinued in 1962 upon the induction of the Schilka weapon system, ZU-23, a 1960 vintage light gun is still going strong, having seen widespread use in the Libyan civil war in 2011 and the current Syrian civil war. The same is also true for ZSU 57-2 the SP anti-aircraft (AA) weapon of 1955 vintage.

Russian ZSU 37 Self-propelled Weapon System

- Chinese ZPU-1/2/4 2 AA gun of 1949 vintage is still in service in China and in over 50 countries around the world having seen action from the Korean war to the Syrian civil war.
- The Pakistan Army Air Defence still operates the 1962 vintage self-propelled weapon system and the 34-year old Artemis gun system of Greek origin.

Compulsions to sustain an obsolete and a vintage inventory is thus a world-wide phenomenon. In fact, no gun calibre the world over has ever been thrown away completely. In that, Sweden, Italy and Singapore boast of their 40mm machines, M/s Rheinmettal is up and about with its 35mm, Germany, Greece and Italy perfect their 30mm, the Russian block with

Poland, Finland and Belgium are going strong with 23mm and the French, USA, Israel and Korea are going strong with their 20mm caliber.

The Challenge of Sustaining Legacy Weapons

Sustenance of legacy weapons is ridden with multiple challenges. The same has rolled out in our scenario as under:

- The production lines of the mainframe weapon having long closed down by the original equipment manufacturer (OEM), nothing fresh has fetched up over the years.
- Invariably, in yesteryears, since the weapons were generally bought as whole consignments, (mostly on a govt-to-govt agreement), hardly anyone of them ever came with dedicated simulators. Consequently, decades of operations and training has been conducted on the war equipment itself. Resulting in every single piece of old weapon going far beyond its designed operational hours/ service life
- With the finite spares that came with the original consignment or the quantity that kept coming till such time the production lines were alive, could hardly suffice to keep the equipment going for decades of extended life. This phenomenon played out as under:
  - After the initial years of sufficiency, the paucity started showing its ugly face (around the mid-life overhaul). This kept becoming worse year after year, not only diluting the quality of overhaul, but also making the day-to-day sustenance increasingly difficult.
  - Also, since in yesteryears, most legacy weapons came without a lifetime-buy (LTB) of spares clause, these measures got applied later as the situation became more and more critical and equipment sustenance started posing an existential threat.
  - With only a trickle of spares that actually arrived in a sporadic mode from the LTB route also, following fait accompli measures started to get applied:
    - Cannibalisation from a part of the inventory to keep the balance alive.

- Forgoing the overhaul option and only resorting to 'optimisation,' i.e., minimum maintenance to keep the equipment from becoming out of action (EOA).

- The surface-to-air missiles (SAMs) pose a totally different set of problems, that is, of shelf-life. What to do when the OEM specified shelf-life expires. Obviously, shelf life extension. How many? One, two, three, after which the issue starts becoming critical as the perishable parts simply start to age, the electronics components start losing their reliability and not much can be said about the continued efficacy and lethality of the warhead lying in passive compact mode for long periods of time.
- Also, with legacy missiles, there are huge problems of reference standards that need to be applied after so many years of storage. Do the original reference tables apply? Are the tolerances and margins in the said tables to be kept unchanged or are they to be modified? If the latter, how much leeway exists and, more importantly, what is the logic? These questions require specific answers which are not always available.

L 70 Towed Gun

ZU 23-2 Towed Gun

ZSU 23 -4B1 Schilka Weapon System

Strela 2M Missile

Igla 1M Missile

Strela 10 M Weapon System

OSA AK Weapon System

SAM 6 Weapon System

The vintage inventory of ground-based AD weapons held by the Corps Army Air Defence

These are the types of grave challenges that stare the air defence warrior in the face. A possible way out is suggested at the succeeding paragraphs.

A Possible Way Out

As regards the towed mainframe gun systems, very 'enabling upgrades' are possible that give a lease of operational life to the weapon. A typical L 70 gun upgrade involves removal of the vintage target laying systems and replacing them with an electro optical fire control system (EOFCS), complete with its day camera, night camera, laser range finder and fire control computer.

Since the vintage ordnance cannot be retro-modified to receive the type of the cutting edge ammunition of today, which has built-in features of receiving round-to-round intelligence of instant muzzle velocity and resultant time-of-flight besides a palletised sub-munitions filling along with intelligent proximity-fusing for maximising the lethality and accuracy at the target end, optimisation of the existing ammunition is the way forward. In that, the conventional single charge pre-fragmented high explosive/tracer

(HE/HE-T) rounds as well as the armour-piercing rounds are replaced with rounds carrying preformed fragmented cubes (PFFC) of tungsten material coupled with proximity fusing. Experience has it that PFFC firing is about 60-70 per cent more accurate than the erstwhile conventional ammunition.

Also, old vintage guns have limitations up to which their rate of fire can be ramped up. Notwithstanding this, with modification in the buffer design and reducing the length of moving part trail sequence, it has been possible to reach higher rates of fire.

In addition to the above, in self-propelled guns the vintage analogue radars can be replaced by state-of-the-art digital phased array radars. These systems are not only more compact but also far more accurate in target tracking. Replacement of erstwhile fuel guzzling gas turbine engines (GTEs) with fuel-efficient digital power packs is another smart measure that can enhance the power and mobility of the old war horses.

Another challenge faced by legacy systems is night blindness. This void is slowly getting diluted as thermal imager/forward looking infra red (FLIR)-based night sights are getting into place slowly.

The near absence of the supply of spares of legacy equipment is a major issue. Since the spares support in range and depth keeps drying up as the weapon ages, indigenisation of spares to the extent possible by any means including reverse engineering is the way to go. Such an effort is the hope of the users. The challenge is indeed colossal.

SAMs, beyond the designated/extended shelf life, are a reality to be faced. This challenge can be addressed along several routes as indicated below:

- Short life extensions (yearly) based on detailed intrusive test of lots with large sample sizes.
- Refurbishment of perished rubber parts where available.
- Re-ploughing spares by cannibalising some of the beyond-repair lots to keep alive the balance inventory.
- Trying to develop reference standards through actual field tests of 'as is' lots.

Another practical approach is to phase out the total quantity order in small packages (as operationally permissible) over an extended timeframe. This will keep the shelf life alive for much longer periods. For this to happen, the immediate requirement could come along with the mainframe weapon system while the balance could be spread over a much longer period.

Finally a few points based on experience:

- Simulators along with the main equipment are a must. Operational hours of the mainframe weapon systems will become more and more dearer as the years go by.
- LTB provision has to be a part of initial negotiations, not a crises-afterthought. The desire to keep it out initially (probably to keep costs low) ultimately results in paying many times over at a later date.
- Timeframe for the OEM to keep spare lines open for an envisaged time schedule (plus stretch factor) must be negotiated at the time of initial contract itself.
- Indigenisation effort for spares must start early in the life cycle of the equipment and not when the situation becomes a crisis.
- In cases where maintenance transfer of technology (MToT) is being negotiated, the following must be ensured:
    - Like all ToTs, MToT is also a mind game. It does not happen 'on file' but by total hand-holding till complete indigenisation is achieved. Co-development/co-production/buy-back are key words.
    - The maintainer must be taken beyond build-to-print (BTP) and build-to-specification (BTS) to the 'know why' in specific areas of maintenance.
    - The normal tendency of the OEM to keep that last bit of ToT up the sleeve for perpetual dependence has to be fought through and eliminated (easier said than done).

These are the dynamics of the challenges to keep alive an old vintage and obsolete inventory. The good thing is that most of these are underway in countering the challenge of 97 per cent.

CHAPTER 9

Simulators in Ground Based Air Defence (GBAD) Ground

"Simulators are amazing—they create reality where none exists"

Brig MKK Iyer, (Retd), SM

Simulators are great enablers and force multipliers. They are not only real-life training facilitators, but also, are a platform for dream-playing for designers of future combat/sensor platforms. In the field of air defence, such simulators (one example is EADSIM)[1] provide many-on-many theatre-level simulations of air and missile defence warfare which act as an integrated analysis tool for experts to model the performance and predict the effectiveness of ground based air defence weapons (GBADWS), SAMs, ballistic missiles, cruise missiles, aircraft and more, in a variety of simulated operational scenarios. Besides, such simulators also serve as realistic air defence training aids.

Simulators are particularly relevant in the GBAD scenario for many reasons. Most obvious of them is the space constraint. For instance, in any 'cabin based' GBADWS, be it a sensor, shooter or a battle management system, there is hardly any space to accommodate anything more than two/three crew members for training. If a larger group is to be trained, either the operational equipment has to be flogged again and again (which will quickly eat up the very finite operational hours) or alternatives need to be found.

In addition to individual training, any real-life air defence field training event involving a group of air defenders demands tremendous infrastructure and resource requirements to be placed on ground. This includes surveillance and fire control equipment, gun and missile systems, command and control systems and more. To work up all these, live aircraft sorties are required to generate a realistic air threat environment before any field training can be imparted to the crew. Organising training in this manner not only comes at an exorbitant cost, but also the same is essentially a one-pass event (implying one time training). This training having been played out once in real-time cannot be recreated (implying rewinding and re-playing) again and again. For every additional training event, not only all the costs have to be re-incurred, but also, the operational equipment must lose another cycle of its operational life.

Post the conduct of the training event comes the next challenge of recording and analyzing the trainee performance. This itself is a very complex exercise, given the fact that the entire air defence training cycle is just about a few fleeting minutes end-to-end. The challenge, very briefly stated, is of locating the fast-moving air threat vehicle, working up the weapons systems and launching of munitions in beyond visual range (BVR) engagements before the air threat vehicle has a chance to release its lethal load. This complete sequence must not only be completed in a few minutes, but also recorded accurately second-to-second for analysing successive trainee performance in each part of the engagement cycle. Doing all this and more with real weapons and equipment is almost unimaginable. Simulators are thus a necessity especially in the GBAD scenario. These can create reality where none exists and can bring realism in training of not only individuals, but also, the entire group at one go incurring comparatively minuscule cost.[2]

Based on the multifarious training requirements briefly stated above, air defence simulators come in various forms depending upon the levels of sophistication, degrees of automation and measure of technological prowess required. Some salient details in this context are at succeeding paragraphs.

Simulators for Crew-Based Training of Larger Groups/Class

To train an entire group or a class without having to suffer the space constraint, most of the radar majors have produced class-room variants (CRVs) of their radars. These variants feature the actual radars stripped open in realistic operational conditions for the whole group to be trained at one go. The realism in CRV-based training is brought about by signal scenario generators (SSGs) working in consonance. These SSGs emulate a variety of fly-over conditions to include multiple attack profiles, weapon delivery passes, electronic warfare (EW) attacks, etc., providing a realistic air-situation environment for radar crews to practise their skills. SSGs also act as platforms for sensor designers to test their new products/test equipment against multiple attack scenarios.

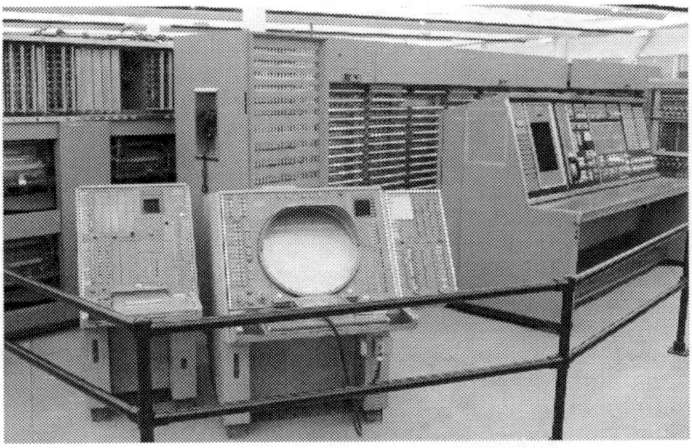

Class room variant of a radar

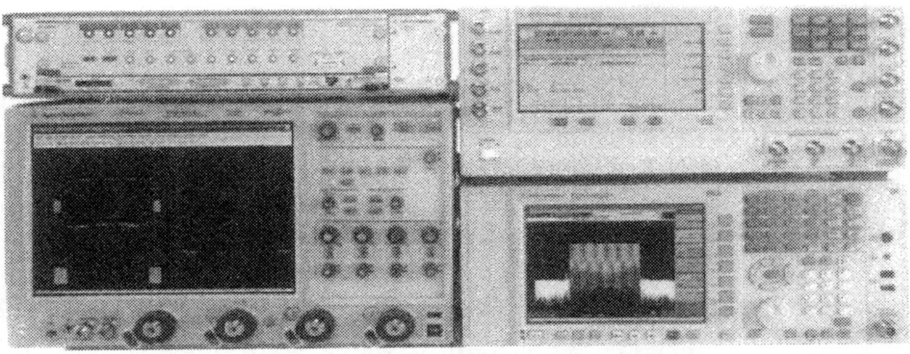

Microwave Signal Scenario Generator

Gun Simulators

Simulators for training the gun crews have different shades of configuration. The basic models generate 3D aerial scenarios on a stand-alone screen on which attack scenarios are simulated. The firer response is accompanied by the actual feel of gun firing (jerks, sound, equipment vibration, etc.). These basically address the training requirements of layer crew on basic sighting systems bringing in as much reality of live firing as possible.

Basic Simulator for L-70 anti aircraft gun

The higher order gun system simulators depict a realistic aerial environment with banks of inter-connected screens or dome-based structures in which simulated air threat vehicles (aircraft, attack helicopters, unmanned vehicles, cruise missiles, anti-radiation missiles, stand-off weapons, smart/intelligent ammunition) execute synergic air threat manoeuvres unfolding to a design and duly separated in time and space, as if in actual.

The threat, as depicted above, is taken on by crews with the help of high-technology simulators that create virtual reality of conventional sighting systems, radar-cued fire-control systems, electro-optical fire-control systems (EOFCS), communication, battle management command and control (BMC2) systems, etc. Such simulators also cater for various ammunition types and their effect on the target end. There are provisions to cater for intangibles like age of equipment, EW suite of aerial threat vehicles etc.[3]

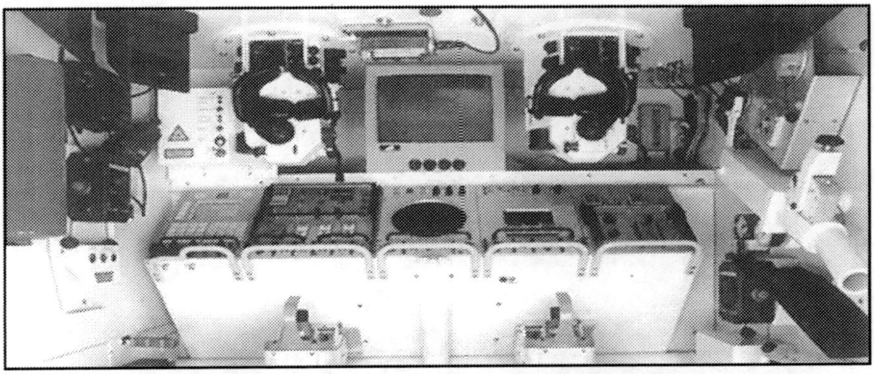

Anti Air Gun Tank Simulator for GEPARD Air Defence (SP) System

Simulators for Very Short Range Air Defence Systems (VSHORADS)

VSHORAD SAM systems comprise mainly comprise of a variety of short-range, man-portable/pedestal mounted missiles. These could be different types like heat-seeking missiles (Igla 1M, lgla-S), laser-guided missiles (RBS-70/Bolide), proximity/impact enabled missiles (Mistral), hit-to-kill with laser-guided end-game missiles (Star Streak), etc. Training of crews on such a wide variety of SAMs requires multiple levels of practices on laying the missile in various air threat conditions, tracking the target steadily, identification of launch zone and a steady launch (which could also be in a hostile EW environment).

To address the above training the challenge, most of the VSHORAD makers offer generic simulator solutions usable for a variety of VSHORADS (for example, the Konus simulator offered by Rosoboronexport of Russia can also be used for training on RBS-70 missiles).[4]

The latest trend is to produce a weapon-effect signature simulator which replicates the launch signature of many types of shoulder-fired SAMs. The

MANPAD Simulator

replication includes the weight/jerk/vibration feel and a self-consuming pyrotechnic that replicates the real effect but leaves no residual projectile. All this is projected against a virtually created environment giving a 3D effect of the launch event.

In such simulators, multiple target movement with attendant visual light and sound clues enable training in evasive action, and counter-action scenarios. Effect enhancers like actual/created weather or time-of-the day etc help in further enhancing realism.[5]

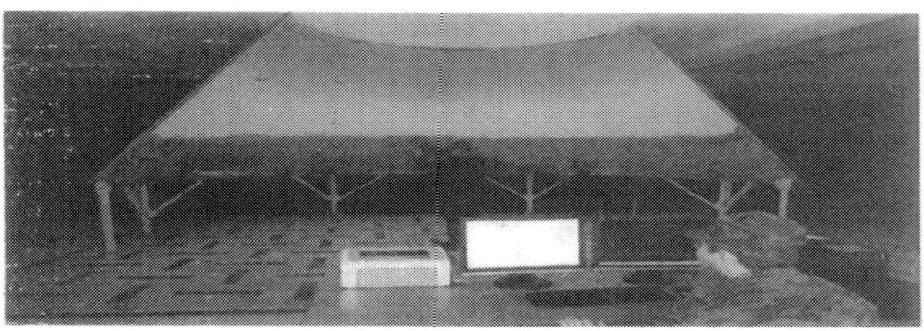

Konus Simulator

Higher-Order Simulators

For higher order SAMs in the short/medium/long category (SRSAM, MRSAM, LRSAM), simulator trends are divided by way of combat functions. For such SAMs, there are stand-alone launcher simulators, driving simulators, sensor simulators, etc. to each one is dedicated to create virtual reality of its relevant combat environment. Typical simulators in this class include motion platform simulators, simulators for commander, operator gunner workplace simulators, simulators for background and target situation replay and simulators for creation of realistic air situation/air threat environments.[6]

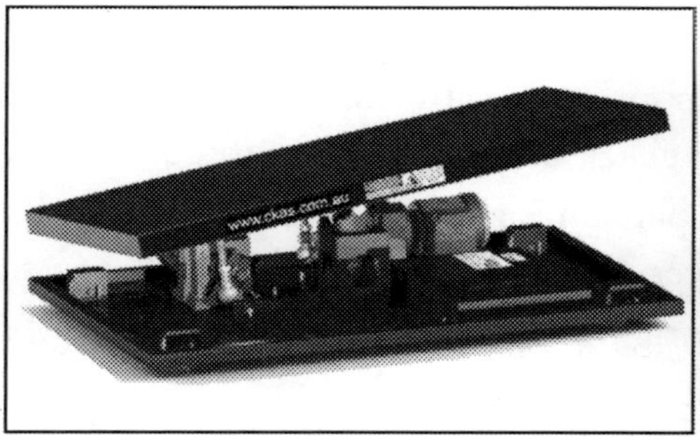

Motion Platform Simulator

Simulation of Commander Operator and Gunner Workplaces Motion Platform Simulator

Background and Target Situation Relay Simulators

Realistic Air Situation Simulator

Geographic Information System (GIS) Enabled Simulators

Today, GIS technologies combined with digital elevation models (DEMs) provide opportunities to evaluate radar beam blockage and other ground clutter phenomenon.[7] These technologies use the potential of the GIS to present topographic information in all its digital detail while specially developed software tools and programmes provide the technical signatures of the radar ordered to the desired range-height matrix as defined by the user.

With these inputs in hand, the system then executes an interplay between the 'technical radar signatures' and the GIS created 'three-dimensional (3D) ground.' The resultant output is the radar coverage diagrams which are accurate to the core.

The great advantage of such an interplay is the inherent flexibility and dynamism provided to the user to check out any number of radar sites for their comparative merit (this comparison is also software driven as well as dynamic) besides having the power to generate real-time changes in the coverage pattern when the radar origin is moved on a mouse-click within the permissible area of deployment, the other variables remaining constant. Good sites/not so good sites/poor sites/overlap, etc., become eminently visible and hence exploitable. This was unimaginable on 2D static map-based display.[8]

Weapon Envelope Visualizers

The erstwhile method of generating weapon envelopes and weapon deployment choices in order to address a particular air threat was an elaborate and a manual process of paper-based planning. This was archaic, besides being time consuming and inflexible. Also, the manual procedure was neither open to dynamic changes nor flexible to comparing multiple deployment choices in real-time.

GIS Enablement[9]

Much of what was earlier stands enabled on the wings of GIS-based technologies and 3D analysis systems. Such systems provide advanced visualization, analysis and surface generation tools that permit viewing large sets of data in three dimensions from multiple viewpoints, ability to query a surface virtue or characteristic and create realistic perspective images that drape raster and vector data in response to the query.

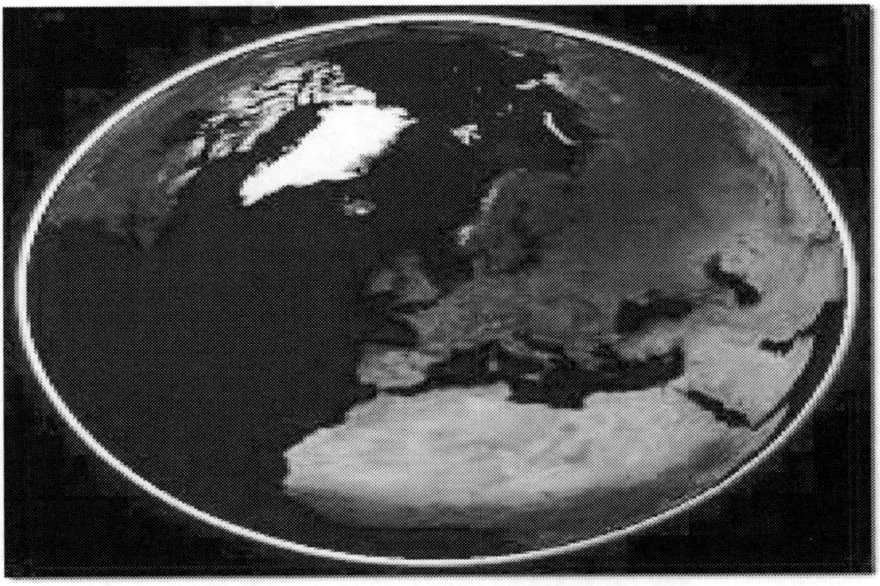

GIS enabled multiple 3D map layers

Exploitation

Exploiting such technologies, 3D virtual maps can be created of the type of terrain over which training is to be imparted. On such terrain, the technical dimensions of the air threat to include technical nuances like range, height, weapon type (conventional/PGMs), stand-off range, flight path details, etc., as well as the technical prowess of GBADWS in terms of range, height, type of ammunition, kill effectiveness, etc., are interplayed by the system in 'one-on-one' and 'one-on-many' modes. This interplay provides effectiveness details of the GBADWS envelopes in taking on the threat, as defined. It also provides a comparison tool to check out, compare and optimise deployment details through a change in weapon type or its location in real-time.

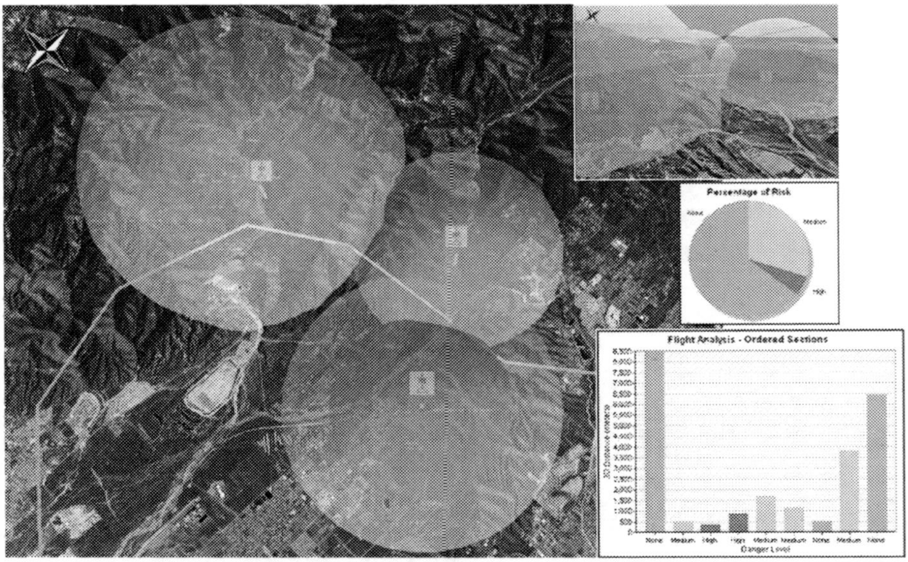

Nuances of air threat vs capabilities of GBADWS interplayed by simulators

Total Effect Simulation

As the technology moves forward, there is a progression towards simulating total air defence battle scenarios. Such systems work on two-side simulation, i.e., both from the attacker as well as the defender. For the former, the system will input various capabilities of the simulated aerial threat vehicles. These may include range, throw-weight, ammunition effectiveness, day/night operation limit, EW muscle, etc. For the defender, the system will input various air and ground-based air defence weapon capabilities in the form of their technical signature and effectiveness against the threat.

The simulator is then played with the defenders utilising their sensors, combat and command and control means in a synergetic fashion. The effectiveness of the defender in optimising their weapons to ward off the 'threat as a package' is accurately calculated and re-playable.[10]

Deployment Simulators

Simulators are also active in the field of dynamically generating deployment options for a set of GBADWS against a defined air threat. Such simulators require inputting of threat definitions in terms of range, height, weapon release line, stand-off capability, day/night effectiveness and munitions on one side, with the technical signature of GBADWS on the other. In addition, effectiveness/success criteria of combat is also to be defined. Pitching one against the other, the simulator generates several deployment options in accordance with the terms of reference of the criteria. These can be compared and contrasted before actually proceeding for deployment

Simulator aided selection of suitable deployment site test

Radar Coverage Prediction Simulator

Based on the inputted technical signature (range, peak power, pulse width) of a radar along with the details of the 3D terrain where such a radar is to be deployed, such simulators generate radar coverage prediction details using the tools of virtual reality and GIS integration along with an applied mathematical model.[11] As their output, such simulators generate multiple radar deployment options each optimising conditions as defined by the user who makes the final deployment call.

Radar Coverage Prediction Simulator

Aircraft Recognition (ACR) Simulation

ACR to distinguish between own and enemy aircraft is an important combat function requirement for air defence warriors. The latest simulators in this field make use of virtual reality to create 3D terrain over which simulated combat (true-to-actual) models of aircraft/aerial threat vehicles are pitched in unlimited and unpredictable profiles presenting real-life scenario for the trainees to recognise them in motion. The entire recognition exercise is played over virtually created environments depicting terrain, atmospheric conditions, timing of the attack, special effects like rain, storm, cloud base, snow, etc. The feel is as good as real.

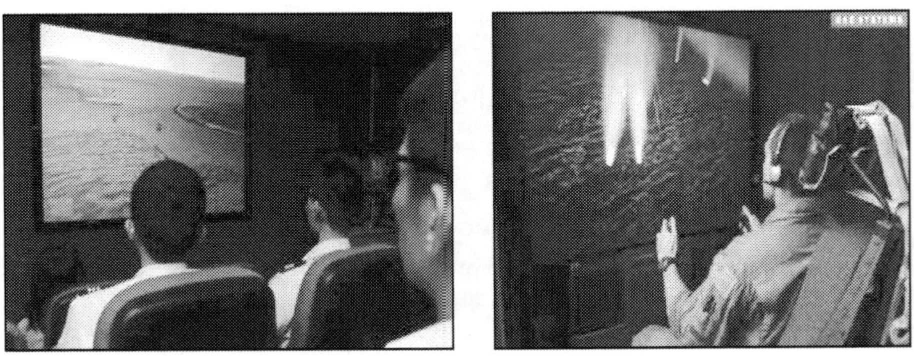

Aircraft recognition simulators

Electronic Warfare (EW) Simulators

Given the growing severity, lethality and diversity of the EW capabilities of the aerial threat vehicle, one of the greatest challenges is to train radar operators in a hostile EW environment. Since realistic field training in this domain is multi-disciplinary hence very difficult and cost prohibitive to implement; simulators as 'training facilitators' are a boon. Such simulators generate an unpredictable regime of electronic counter measure (ECM) attacks to which the victim radars are subjected. Operators are then trained to work through these attacks and keep the sensor on the target. Further advancement in this field is a two-way simulator in which not only the EW attacker but also the victim radar is simulated. This makes the entire EW training package executable in a class room.[12] In another variation, the ground-based jammer could also be on an airborne platform, or in the field.

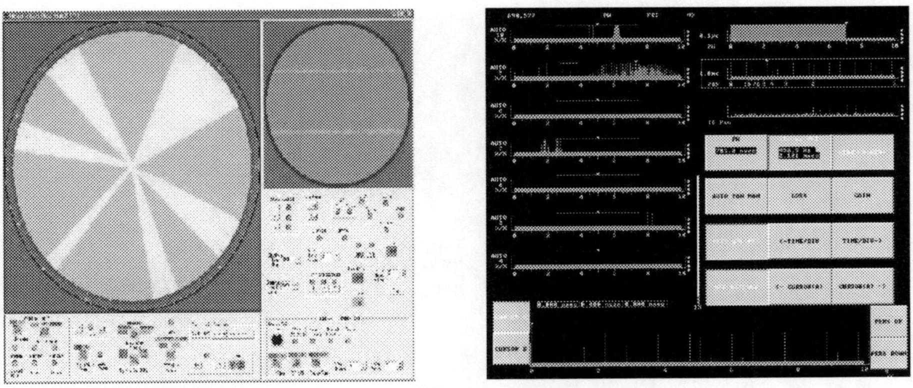

Electronic Warfare Simulator

Such is the exciting and ever-evolving world of simulators that hold a special place of significance in the GBAD scenario. In essence, riding on the wings of cutting-edge technologies, the enabling power of simulators must be exploited in a big way by GBASWS to create realism in the training of air defence crews.

NOTES

1 https://www/eadsim/com/. Accessed on 09 Jan 2018.
2 https:// www.elbitsystems.com/Elbit main/files/Air Defence.pdf. Accessed on 09 Jan 2018.

3 www.kmweg.de/49655-bD!lbg-PRODUKTF-Training. Accessed on 09 Jan 2018.

4 Worldwide-defence.blogspot.com/.../sa-24-grinch_Igla_S-manpad-data. Accessed on 09 Jan 2018.

5 www.flightglobal.com>News>Space>Manned space flight. Accessed on 09 Jan 2018.

6 Simulator.com.ua/files/tp/Tunguska_en.pdf. Accessed on 09 Jan 2018.

7 [HTML] Radar Beam Occultation Studies Using GIS and DEM Technology: An Example Study of Guam [HTML], PA Kucera, WF Krajewski-journals.ametsoc.org.Accessed on 09 Jan 2018.

8 Military Applications of GIS.www.gisdevelopment.net/application/military.../militaryf0002pf.htm. Accessed on 09 Jan 2018.

9 Arc GIS 3D Analyst.www.esri.com/software/arcgis/extension. Accessed on 09 Jan 2018.

10 www.defencesindustrydaily.com/ 101m-for-air-defence_simulators. Accessed on 09 Jan 2018.

11 www.ieeexplore.ieee.orWie15/5482534/5494381/05494627.pdf. Accessed on 09 Jan 2018.

12 Google images: aircraft recognition simulators attachment.pdf. Accessed on 09 Jan 2018.

CHAPTER 10

Future Air Defence Needs

Lt Gen (Dr) V K Saxena, (Retd), PVSM, AVSM, VSM

Having analysed the multiple nuances of air defence battle and various other contemporary issues related to the vast domain of ground-based air defence (GBAD), this chapter attempts to spell out the future GBAD needs at a macro level. This has been done by drawing out from each of the major issues discussed in the foregoing chapters. Since the entire content is open source, the needs have been spelt out in generic and conceptual terms.

Ab-initio, it needs to be stated that the tendency to see the air defence resources in service-specific (Army, Navy, Air Force,) confines and verticals must be guarded against. In fact, all the ground-based air defence weapon systems (GBADWS) of the country, composed of a trio of sensors, shooters and battle management systems, must be seen as one cumulative national resource pitched as a whole to counter the air threat posed by our potential adversary(ies).

The next issue relates to the definition of the air threat in qualitative and quantitative terms. "It is the air threat that drives air defence" is the fundamental position, as the two are tied to each other in an eternal cause-effect relationship. Suffice to say therefore, that if the air defences are to be kept razor-sharp, the threat which it must counter must also be kept always up to date at all times. This exercise of updating the air threat at regular intervals through an institutional arrangement must continue. (mode of execution not discussed, being classified).

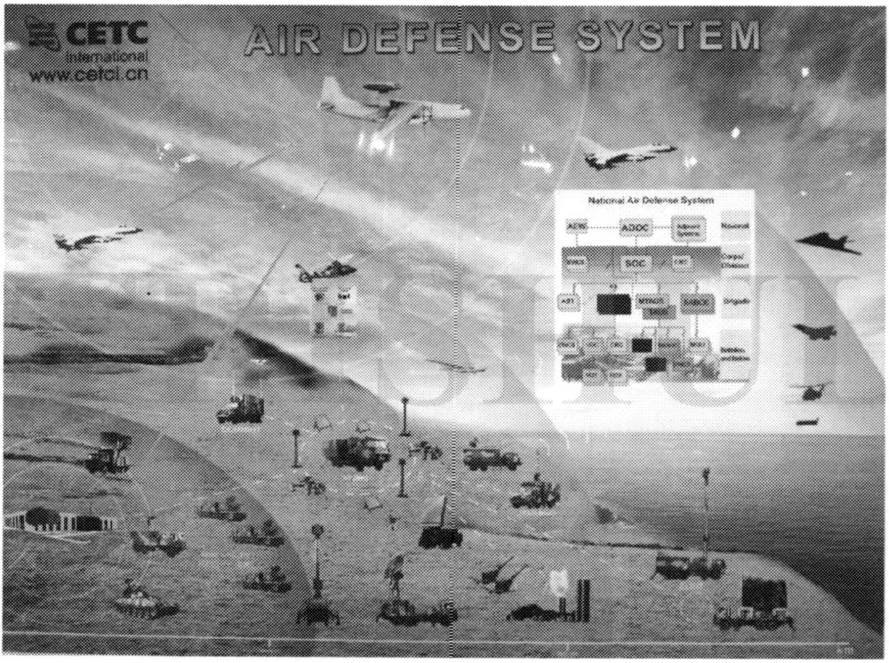

Chinese Integrated Air Defence System

Having stated the primary requirement of updated air threat being made available to the stake holders from time-to-time, the next logical step is to define what vulnerable areas/vulnerable points (VAs/VPs) need to be provided air defence against the defined threat and with what range and depth of GBADWS. VA/VPs imply those assets which merit provision of air defence cover to protect them from adversary's air attacks. Dimension wise, VP is up to 1000×1000m while the VA is 3000×3000 m).

For the above, an institutional framework is already in place in the form of a joint services study group (JSSG). The details of this JSSG are not discussed as they are classified. It is imperative that this JSSG is commissioned at regular intervals so that a prioritised list of VAs/VPs, not only of the Services but also relating to civil/strategic domain is made available along with the details of the type and quantum of GBADWS required to provide them air defence cover. Actually, the deliberations of the JSSG lay out a roadmap for procurement of GBADWS over various

Plan periods. These get suitably incorporated in the various perspective planning documents driving defence procurements.

While the specific details of the JSSG are beyond the purview of this book 'essence points' in generic terms are stated:

- To effectively counter the contemporary as well as futuristic air threat, an integrated family of GBADWS is required consisting of guns and a hierarchy of SAMs starting from very short range to short range to medium and long range.
- A combination of sensors (surveillance and fire control systems), shooters (guns and missiles) and a fully-packaged battle management system (capable of seamlessly integrating similar and dissimilar systems) is required as an integrated air defence system (IADS) which also includes, besides the GBAD, the air defence means in the air and at sea (last two not discussed).

Flowing from the above, what are our specific future needs of GBADWS and what challenges need to be overcome in realising each of these is covered in the following paragraphs.

Gun Systems

L 70

Starting with the mainstay gun system, i.e., L 70, this weapon was inducted in the Indian Army at the beginning of 1963.[1] The weapon has been kept operational with repeated overhauls in a spares-starved environment.

The L 70 guns have been put through a comprehensive upgrade so as to give them an operationally viable lease of life covering the period in which the successor weapon can be realised. Not only for covering the period for the successor gun, an upgrade of the existing weapon system is also a norm the world over as no country ever throws away a calibre.

The details of this upgrade have already been covered in chapter 7.

The number of guns to be upgraded has been worked out in detail keeping in mind a host of factors such as the vintage of the chassis to take

L 70 Upgrade

the upgrade, likely de-induction schedule of un-upgraded guns coupled with the likely phased induction of successor systems and the likely lease of additional operational life after the upgrade. The critical factors in this calculation are two. Firstly, the bottom-line requirement of the total number of guns must always remain operations ally available and, secondly, this bottom-line number must at all times be met by a combined pack of upgraded+un-upgraded guns till de-inducted + successor guns. In fact, the de-induction schedule has to be very finely balanced with the induction schedule keeping a positive overlap and always ensuring that the net holdings at any point in time never fall below the said bottom-line requirement.

This number duly worked out is known to the stake holders. The same is not quoted here as it is classified.

The following points are stated as to this upgrade:

- As stated, the numbers to be upgraded are extremely critical. They must be adhered to. In case the throughput of the upgrade lags behind in time and space for any reason whatsoever the total quantum of guns to be upgraded will have to be re-calculated again.

- The original equipment manufacturer or OEM (in this case implying the agency doing the upgrade) must ensure that the inevitable delays of the teething trouble period are made up by enhanced throughput now, as getting upgraded numbers in a time critical sequence is planned carefully and impacts net operational effectiveness. The proof of successful upgrade is live firing.
- There must be a total hand-holding by the OEM even after the upgraded guns are handed over to the user till the time the latter become fully confident to take it ahead alone.
- Not only the weapon upgrade per se is important, it is the continuous supply of spares that must be maintained by the OEM right up to the life cycle of the upgraded weapon system. It is also to be noted by the OEM that during the initial teething trouble period, when the upgraded weapon system actually settles down in the hands of the users, there will be a much higher requirement of spares. The OEM has to cater for this surge.
- Guns without adequate ammunition are a trigger-less arsenal. The L 70 weapon system uses both conventional single charge HE/HE-T and armour-piercing rounds besides the modern preformed fragmented cubes (PFFC). It must be ensured that an adequate quantum of ammunition in terms of First Line, Second Line and a War Wastage Reserve (WWR) for a designated intense period (not specified, being classified) both for the upgraded as well as the un-upgraded guns is made available by the authorities concerned

ZU 23-2 Guns

Zu 23-2 fair-weather twin guns are another mainstay gun system held with Army Air Defence. ZU 23 guns were inducted in the Indian Army in 1973. The weapon has been kept alive and operational by successive overhauls in a spares-deficit environment.

As per the open source, M/s Punj Lloyd is in the final stages of signing a contract for Rs. 670 crore for upgrading 468 ZU 23-2 guns. For this upgrade, M/s Punj Lloyd is to partner with a Slovakian defence company,

EVPU. This end-state has been reached after a due procurement process sequence which involved multi vendors.[3]

The main constituents of the upgrade run on similar lines as that of L 70 gun upgrade. This basically includes an incorporation of an electro optical fire control system comprising of a day camera, a night camera, a laser range finder and a fire control computer.

Operational considerations such as the number of guns to be upgraded, the quantum of ammunition required, the need for hand holding in the initial teething trouble phase, requirement of spares for the life cycle of the upgrade (normally assessed as 15 years) and the need to cater for the surge requirement of the same are similar as explained for the L70 gun.

As regards the ammunition for ZU 23-2 guns, it is to be ensured that not only the quantum as required by first line, second line and WWR to fight the specified days of intense war (I) is made up (exact quantities not quoted being classified), it is very important to ensure that the ammunition is safe to fire.

Also, the urgent operational need to sign the contract cannot be over-emphasised. Any further delay may result in a situation where the un-upgraded guns are no more sustainable while the successor guns have still not arrived.

Successor Gun System

Any amount of upgrade of the existing guns cannot wipe away the crying need for a successor gun. This is so because the upgrade per se, can only give a limited lease of life to the weapon system at nearly the same technological threshold or at a little stretch at the best. Successor systems not only bring a new lease of operational life but also, and more importantly, these bring a paradigm jump in the technological level as compared to the 'in-service' systems.

In the context of guns, these manifest as enhanced rate of fire, higher precision in strike, higher commutative kill probability (CKP), longer operational life, minimal crew, lesser requirement of spares, longer time between successive overhauls and a general ease of maintenance, repair and operations.

The above outcomes are brought about by the dual impact of superior design and high technology ammunition. The emerging design features include gas-operated twin barrels with positive lock, faster ramming and ejection through a shortened ramming route, innovative designs bringing minimalistic movement of moving parts over shorter distances, controlled move of running out barrels telescoping several actions in the same time sequence, use of brushless and hydro-pneumatic motors driving auto feed devices and more.

As to ammunition, the trend is for intelligent ammunition. This type of ammunition can absorb the round-to-round instant muzzle velocity and can set the fuse dynamically for a corrected time of flight for high accuracy. In addition, such ammunition features better propellant chemistry which ensures a higher kill effect by itself and through additional features like sub-projectile fillings and their optimal distribution at the target end (example AHEAD or advance hit efficiency air defence ammunition from M/s Rheinmetall Air Defence AG).

It is unfortunate that the case for successor guns for Army Air Defence has been going on for a long time without success. The ups and downs of the case and the multiple reasons for delay are not covered here except the fact that the RFP issued for 428 successor guns at a cost of Rs. 16,900 crore under the Buy and Make category stands retracted.[4]

At the same time, a case for 244 guns (also referred to as close in weapon systems or CIWS), 228 fire control (FC) radars and 204,000 programmable ammunition for the Air Force at a cost of USD 1.5 bn is under progress. RFP for the same stands issued under 'Buy and Make (Indian)' category.[5]

Following points are stated in the context of successor case:

- Procurement of successor guns is not a matter of choice but an issue of operational expediency on which rides the core capability of air defence forces to ward off the enemy's air threat at the terminal end.

- Notwithstanding the open source, the exact number of guns required as well as the specific types and quantum of ammunition stands mentioned in the respective procurement cases of the two services. They must be procured without any further delay.
- The stalled Army's case, post the retraction of the RFP, must be re-launched post haste either on a tri-Service format or stand-alone.

Self-Propelled Gun System

As stated earlier in chapters 1 and 2, self-propelled air defence gun systems are required to provide close air defence cover to mobile assets of the field force. Army AD has two types of such systems, namely, the vintage ZSU23-4B1/M Schilka weapon system and the 2k22 Tunguka gun missile system (quantum holding not mentioned).

Two cases are going on in this field. One is the upgrade of the vintage Schilka system[6] and, secondly, the procurement case for the successor to Schilka and Tunguska weapon systems.

The Upgraded Schilka Weapon System

Following imperatives are stated as to these cases:

- The quantities both for the upgrade as well as for the successor systems stand specifically mentioned in the respective procurement cases. The same are not quoted here.

- All the points stated for the OEM in the case for the L 70 upgrade (time lines, spares, responsibilities, et al) are applicable in the case of the Schilka upgrade as well. The spares issues in the case of the Schilka system are much more critical due to their near non-availability status the world over.
- The OEM doing the upgrade is also to ensure that the overhaul of the system prior to upgrade is ensured in letter and spirit.
- Since the upgrade involves the collaboration with a foreign OEM, it is most important to ensure that there is no perpetual dependency on the said foreign OEM. This can only happen if the continuous system support is taken over by the Indian partner to the contract sooner than later. It is heartening to see that the same has already started to happen. The bottom line of acceptance by the user only after live firing must remain the litmus test.
- The successor case relates to a critical void in the existing arsenal. This must be progressed as fast as possible. The current quantum of systems taken for upgrade are based on a calculation of the time frame for the successor. In case the same exceeds the planned figures, a re-think on the numbers for upgrade will have to be gone into. This contingency already stands worked. Exact numbers are not quoted here.

Missile Systems

As explained in preceding chapters, besides the towed and self-propelled guns, the balance of the integrated family of GBADWS comprises of missiles starting with VSHORADS (man-portable, shoulder/pedestal fired very short range missiles) which combine with guns to provide the gun-missile mix deployment with a coverage up to 6-10 km, short range SAMs (SRSAMs) providing air defence missile cover up to 20-30 km, medium range SAMs (MRSAMs) up to 70-100 km and long range SAMs (LRSAMs) for more than 100 km. Each of these missiles is discussed at succeeding paragraphs.

VSHORADS

Like all GBADWS in the integrated family are important in their own right, so are the VSHORDS. The requirement of this weapon system is basically to provide a gun-missile mix at the terminal end. This requirement is inescapable as a single-tier defence by terminal AD guns alone is considered highly inadequate for providing comprehensive cover at the last line of defence in countering the contemporary, as well as, the futuristic air threat.

It must be appreciated that the terminal end actually signifies 'the last line of defence.' This line will actually face a determined/smart/techno-savvy attacker who has either successfully penetrated and survived in transiting through the multiple layers of missile defence based on its technological muscle of EW, stealth, or superior avionics and more, or has made a surprise attack taking advantage of other factors like terrain and (or) weather.

The VHORADS alongside the terminal guns at the last line of defence provide the basic minimum quantum of fire to take on the determined attacker in the proverbial 'last man last round.' As the severity and lethality of the air threat continues to be revamped, the need for the effectiveness of this last line defence becomes more and more critical.

Besides just enriching the quantum of fire, the VSHORADS in a gun-missile mix also provide the following:

- Provide the capability of dynamic deployment wherein a defender can continue to switch between different options of combining guns and missiles, be unpredictable and be ready to surprise an attacker by bringing effective fire on it from a totally unexpected position/direction.
- In certain critical deployments, especially in mountainous/high altitude or glaciated terrain, an asset being protected can be approached only from a particular direction(s). VSHORADS in these cases greatly help in increasing range, reach and kill effectiveness along the said possible avenues of approach.

- Most of the VSHORADS, being quick reaction (fire-and forget heat seeking/beam riding, et al), prove to be ideal for suddenly appearing threats—typical of attack helicopters (AHs) following the nap of the Earth (NOE) technique of attack.
- VSORADS provide avenues to the defenders to be innovative and unpredictable in the following ways:
  - Deploying missiles way out of the gun ranges thus increasing kill envelopes along likely avenues, albeit with certain gaps at known and acceptable volumes of space.
  - Laying air ambushes along 'inescapable avenues of approach' that pass through narrow valleys and steep passes thus increasing the overall kill effectiveness in unconventional deployments.
- Normally the VSHORDS bring in their own surveillance resources for target early warning and fire control. These add to the sensor capability of radar-controlled gun layouts or ends up in providing the essential 'eyes and ears' to the gun missile mix, where the gun component is fair-weather.
- In a typical and a widely practiced attack technique of the attacker adopting SEAD tactics and suppressing the fire control radars of terminal guns, the VSHORADS sensors normally operating at different frequency signatures provide a way out and a possibility of returning effective fire.
- A similar way out is provided by VSHORADS when the sheer quantum and multiplicity of attack vehicles surpasses the target handling capability of the gun layouts.

In short, the inescapability of a gun-missile mix at the terminal end just cannot be over-emphasised. Appreciating this operational need, the future WE (War Establishment) of gun-alone regiments (Air Defence Regiment) is gun-missile (GM Regiment).

The current VSHORAD procurement case involves three vendors, namely, MBDA (France), Rosoboronexport (Russia) and SAAB (Sweden) with their Mistral 2, Igla S and RBS 70 NG VSHORADS, respectively. The same is going on since 2010.[7]

MBDA Mistral

IGLA S Russia

RBS 70 NG Sweden

Without going into the period 2010-2018, the end state as reported by the open source is all the three vendors are in the final selection for the lowest bid. It is a huge tri-Service order of 5,175 missiles and 1,276 single and multi launchers at the cost of approx USD 5 bn.[8]

It is the hope of the author that:

- That a running case of eight years will move towards contract as a matter of urgent operational necessity.
- That associated aspects of transfer of technology (ToT) to a defence public sector undertaking (DPSU) or a DPSU and a private player in the public-private partnership (PPP) model will come through.
- That the issue of offset obligations along with the challenge of discharge through the Indian offset partners (IOPs) will not become a show-stopper.

SRSAMs

As explained in the previous chapters, SRSAMs are critically required to provide short-range missile cover to our vulnerabilities. The total requirement of SRSAM stands stated in the JSSG and respective procurement documents (not quoted here).

As regards SRSAMs, there are basically two options; the indigenous Akash weapon system or the SRSAMs ex foreign vendors like BAMSE (M/s SAAB, Sweden), TOR M2K (Russia), VL MICA (M/s MBDA, France), VTK Mk 1 (M/s Thales, France) and Spyder (M/s Rafael, Israel) and others. Since the indigenous Akash SRSAM stands successfully inducted into the Services, it is the sense of the author that the future SRSAMs could also follow the AKASH route.[9]

Akash Weapon System

Coming to Akash, while the initial induction is a success story, the challenge for the OEMs is to ensure the realisation of the weapon system as full-fledged regiments get into top gear by cutting out delays in delivery schedules and spares support. It has to be understood that the requirement of SRSAM is critical to provide short-range missile cover to strategic vulnerabilities.

A clear time plan must emerge duly accepted by the OEMs and approved by the ministry of defence (MoD) in which the requirement of SRSAMs for the three Services as specified in the procurement cases (numbers not mentioned, being classified) will be completed.

QRSAMs

Apart from SRSAMs, there is another requirement of providing mobile missile cover to mechanised formations through quick reaction SAMs (QRSAMs). Such weapons are distinct in their own right. They can keep up with mechanised assets, have the unique capability of surveillance and tracking during move and can fire immediately on halt. It is heartening to note from the open source the progress of indigenous QRSAM.[10] The challenge is to take this progress to its fruitful realisation with full involvement and participation of the user and fill the existing void in the QRSAM category for the field force.

In the case of QRSAMs also, the quantum and bottom line requirements of fire units (FUs) also stand specified as explained for SRSAMs. These numbers are not quoted here as they are classified.

For QRSAM also there are two options. Firstly, the indigenous route or, secondly, procurement through foreign vendors. Some QRSAMs in the field are Pantsyr (Russia), Spyder (M/s IAI, Israel) and QRSAM (M/s Rafael, Israel).

Given the fact that the indigenous QRSAM is becoming a success story, it is the opinion of the author that future requirement of QRSAMs should be addressed through the indigenous route.[11]

The critical requirement to commit to numbers and time frame by the OEMs as stated for the SRSAM cannot be over emphasised in the case of QRSAM also. Finally, as a point of caution, it must be ensured that in case all requirements of SAMs get allotted to indigenous OEMs, it must not cumulatively exceed their finite manufacturing capability. Such a situation can result in endless wait for the user besides huge cost overruns.

Having said that, the indigenous QRSAM case must be progressed with all speed.

The indigenous QRSAM

MRSAMs and LRSAMs

As explained in the previous chapters, the requirement for MRSAMs is to provide area air defence cover to our vulnerabilities. They form the upper layer of the layered and tiered air defence cover in an IADS system.

The current status is that all the three Services are on the indigenous route of development of MRSAMs and LRSAMs in collaboration with a foreign vendor.[12] The Defence Acquisition Council (DAC) has approved the number of regiments of MRSAMs for the Army and Air Force and LRSAMs for the Navy under this joint venture (numbers are classified). The requirement for all stake holders is to move forward towards realisation of the system and its early operationalisation. Experience has it that, the users and OEM must not only to look at the weapon vertical alone but also the associated support structures, battle management and command and control spects. Following points are stated as to the procurement of this weapon system:

- Quantum numbers as stated in the procurement documents are the targets to be achieved. The time and cost overruns must be reduced to the minimum possible. In fact, the accelerated production rate must be pressed in to cover the time delay already incurred.

- As stated above, operationalisation does not mean missiles alone but also a whole lot of associated infrastructure for battle management, missile testing, missile storage, spares support, et al. The weapon system must move towards operationalisation as a 'whole' and not as an individual missile vertical.

The MRSAM Joint Venture

Dealing with the Ballistic Missile Threat

The ballistic missile threat has been discussed in detail in Chapter 6. This needs to be countered by a suitable ballistic missile defence (BMD) capability as highlighted in that chapter. The imperatives towards realising this capability are reiterated.

Imperatives

Keeping in mind the net ballistic missile threat in our neighbourhood that exists as of now, the case for BMD for India cannot be over-emphasised. The need of the hour is to move forward expeditiously towards realisation of Programme AD systems as deployable FUs under Strategic Forces Command (SFC).

In that, as the Indian BMD system moves towards operationalisation, certain associated points need serious consideration.

First and foremost, the declared NFU doctrine is essentially based on the premise that we have the capability to take on the enemy's first strike and survive to cause the stated massive punitive retaliation. This would call for building robust and survivable infrastructures, fail-proof and survivable regime of codes and authorisations for weapon delivery, duly supported by several layers of redundant and secure communications. Besides the wherewithal to stage a response, the survivability of actors also needs to be ensured through alternative command posts with a chain of hierarchy fully defined and known to the decision makers up and down the chain.

As for collateral damage and casualties to the civil population, the enemy's first strike is likely to be a calamity, very much like a man-made disaster as explained by the guidelines issued by the Natural Disaster Management Authority (NDMA). These comprehensive and very well documented guidelines which call for a whole lot of actions at the Centre and State levels, need to be "realised on ground" and practiced and rehearsed time and again till the potential victim population becomes fully aware of the looming danger.

Planned dual use of existing infrastructure, like construction of mass shelters, keeping disaster relief bricks at hand, catering for disaster medication and many other calamity management measures as defined and laid down in the NDMA guidelines call for immediate attention and action by the concerned authorities and organisations.

Last, but not the least, will be the requirement to put into place an integrated command control and battle management system which will integrate multifarious 'sensors' and 'shooters' of air defence and anti-missile system into one seamless loop. This life-line system will be responsible for optimising surveillance resources through multi-sensor fusion in the detection of an incoming threat followed by target designation and weapon selection so as to ensure successive punishment to the incoming threat. The basic tenets of these systems are steadily being put into place.

Programme AD and later S 400 systems will have to integrate with the same.

As to the planned operational deployment of Programme AD, open sources reported way back in 2012-13 that once the programme is operationalised (target-2016, likely by 2020-21) the system FUs will be deployed to provide BMD cover to the two metro cities of Delhi and Mumbai. Since the system operationalisation for Phase II was nearly getting complete in 2017, it was reported in *The Times of India* in March 2017 that plans are afoot for deployment of BMD FUs for New Delhi.

The requirement in the immediate will be to ensure that the operationalisation of the system moves forward which essentially implies the following:

1. The system must completely and exclusively change hands 'from the scientists' to the 'users and stakeholders.' In that, the Strategic Forces Command must fully own, operate and actualise the system as a weapon in its arsenal. Full operationalisation must be achieved at the earliest.
2. The sequence of deployment of Programme AD FUs, which involves a large number of actions demanding multi-disciplinary and multi-agency coordination must start to unfold immediately.
3. Sequence of action to realise associated command and control infrastructure, missile testing and storage sites and tuning up the disaster management response must come up alongside

Surveillance and Battle Management Systems

As regards a whole hierarchy of surveillance and fire control (FC) systems, it is reassuring to see that the indigenous capability is coming up reasonably well. The recent inductions of the state-of-the-art radars point to this deduction.[13]

The challenge for the OEMs is to ensure that one, there is total synergy between them, and two, the user is kept on board all the way. Besides this, they must ensure that the mutually agreed completion dates (PDCs) for multiple types of radar under development is not allowed to fall behind.

In fact, the ultimate aim should be to get rid of foreign dependence in this field. While that should be the ultimate goal, the challenge to maintain the current inventory of fire control (FC) and tactical control (TC) radars must not be lost sight of either. This will demand continuous hand-holding and providing spares support to the user.

Besides the indigenous development of sensors, the requirements of the Air Force for filling the sensor voids of IACCS also needs to be addressed.

As regards battle management systems, the challenges of integration of the ADC&R systems of the three Services must be achieved. Also, the technical and obsolescence issues relating to the time frame of development must also get addressed in future systems.

At the cutting edge of GBAD weapon systems, the following is required:

- As the old vintage L 70 gun moves along the up-gradation route, the associated old vintage radars must also be replaced with successor radars.
- The quantum requirement of these radars have been specifically calculated (not quoted here, being classified). It must be targeted in an operationally acceptable time frame.
- Over a period of time, a regime of new-generation radars must fill the existing void in their respective categories (quantum voids known to stake holders) completely replacing the existing old vintage radars.
- Till such time as the above transition takes place, it must be ensured that the existing fleet of radars is not starved of spares. It is understood that it is a tall order but the same is an operational compulsion.

Addressing the UAS Threat

As explained in chapter 5, while the UAS are an 'attacker's delight' they are a 'defender's nightmare.' This is so because not only detecting these vehicles electronically is an uphill task due to their minimalistic radar cross-section signatures (RCS), but also, the challenge is to develop such low-cost weapons to kill them so as to balance the 'cost of kill' with the 'cost of target.'

Launching millions of dollars worth of missiles to bring down low-cost UAS does not make sense.

A suggested way ahead to deal with the UAS threat has been covered at Chapter 5, the same is reiterated:

- As stated above, our basic premise must remain anchored on achieving cost-effective kill on UAS. Bringing them down using close-in-weapons like air defence guns, man-portable missiles or more 'inappropriately' using VSHORADS or SRSAMs or worse by scrambling air defence aircraft in search-n-shoot missions should remain the last option.
- Going with the flow of realised technology at the cutting edge today, we need to go in for two types of cheap kill solutions, namely, 'hit-to-kill' and laser kill.
- Developing a cheap 'hit-to-kill' system of the type of Peregrine Eagle or Cougar is well within the capability of the DRDO with technological and software support from private industry where many a start-ups are already coming of age.
- We already possess the capability of having a basic pilotless target aircraft (BPTA); the missing component is to design a kill vehicle, either strap-on or as a detachable integral that can be put on a guidance course to the UAS. This guidance package must make use of existing sensors keeping the kill vehicle as cheap and simple as possible (no RF/IR seeker to be configured on board).
- The second and parallel direction in which to move simultaneously is to develop the anti-UAS laser kill system. In this field also, excellent work is already at hand (details not covered). This must be taken further with due support from private industry where several players, both indigenous, as well as, foreign OEMs are capable of offering technological capability in the field of lasers. (The foreign OEMs could be taken on the joint venture/memorandum of understanding (JV/MoU) route.
- The user Directorate could provide an integrating platform where the already spelt-out felt need can be addressed by public and private partnership.

- It is relevant to bring out that the cutting edge technology today is already addressing the following erstwhile challenges in the domain of laser kill weapons:
  - Capability to aggregate the power of a number of laser sources at the point of impact so as to generate a kill beam of sufficient intensity.
  - Capability to auto-focus multiple beams from co-located sources at a single chosen point of impact (target) in near real time through an 'auto cued-on' arrangement.
  - Use of optics (gimballed mirrors/reflectors and more) on board aerial kill vehicles to reflect multiple laser beams aggregated from different sources on ground as a strong 'cumulated kill beam' (unlimited magazine) on the target UAS.
- Research in the field of charged particle beam and high power microwave must be kept alive by the DRDO as well as a few other players in the private industry working in this vertical so as to come out in the foreseeable future with possible soft kill solutions in this still evolving field.
- For any new procurement of guns and missile systems, anti-UAS capability (especially the sensor capability to detect minimalistic targets) must be checked out before finalising the procurement.
- Old vintage Russian radars operating in lower frequency bands must not be junked. These are excellent radars that can detect slow moving UAS. Even the sensors on the current holding of GBADWS have shown varying capability of detecting small RCS UAS. These sensors must be used optimally in building cheap kill capabilities (say UAS hit-to-kill system).
- And, lastly, why kill the UAS when the following can be easily done:
  - Make the UAS stream the audio/video data not only to its GCS but also to the hacker station.
  - Make the UAS send deliberately wrong data to its operator.
  - Confuse the UAS with false GPS information, waylay it or direct it in the wrong direction. Just hijack it.

- Take command of the UAV and make it fly to your wish/make it crash/use it like a guided missile. (remote hijacking).
- Extract valuable seed data from the built-in threat library of the UAS and use it to your advantage.
- All these and more are options to be exploited.

Dealing with the Stealth Threat

As enumerated earlier, stealth threat has already arrived at our door. There is a need to develop anti-stealth capability. Relevant action points in dealing with this threat have been covered at Chapter 7. These are reiterated below:

- Since a large number of ground-based air defence weapon systems (GBADWS), spread across sensors, shooters and battle management systems are at various stages of the procurement cycle, it needs to be ensured that all futuristic systems must invariably have the anti stealth muscle.
- Several legacy tactical control radars (those that can electronically designate target tracks to other radars), early warning radars (those that carry out only surveillance with no capability to designate targets to other sensors), fire control radars (controlling the fire of GBADWS) and missile guidance radars of Army Air Defence (exact names not quoted) are gradually on their way out either because their successors have arrived or their product-improved versions are ready. It is to be ensured that incoming radars have a degree of anti-stealth muscle (which does not necessarily means only lower band radars).
- Certain legacy SAM systems of Russian origin held with Army AD still possess the old vintage radars which can detect stealth targets. Even if the main equipment gets obsolete and is to be de-inducted, these anti-stealth gems need to be kept and exploited as long as it is possible.
- Over a period of time, the indigenous capability has gained considerable clout in the design and production of a whole spectrum of radars. It is time for them to turn the focus on radars with anti-stealth capability. A beginning can be made by developing a bi-

static radar. Future R&D areas could be along the path of PCL and LIDAR. In this, it is not necessary to re-invent the wheel; a head start is available along multiple routes of joint ventures/ memorandum of understanding/transfer of technology (JV/MoU/ ToT), etc., with foreign original equipment manufacturers (OEMs) under the Make in India initiative.

- In the context of producing anti-stealth equipment, what applies to the defence public sector is also equally applicable to private industry. Since the stealth muscle in our scenario is just about maturing, it is only going to become a more and more prominent feature of the futuristic air threat. Opportunities to develop stealth detecting radars for indigenous consumption, as well as for the export market either on the 'go alone' mode, or through technological tie-ups, is thus an emerging business opportunity.

The above in essence are the future air defence needs of the country.

NOTES

1 Modernisation of Army Air Defence, Indian Defence Review, Vol 29.3 Jul-Sep 2014.

2 https://www.defenseworld.net>India's-BEL-displays-L-70_Artillery-gun. Accessed on 13 May 2018.

3 https://www.defproac.com>Punj-Lioyd-to-sign-contract-for ZU23-2-B -upgrade. Accessed on 12 May 2018

4 https://www.financialexpress.com>army-set-to-issu-rfp-for-16000-Cr-gun-programme. Accessed on 12 May 2018.

5 https://www.defensenews.com>india-to-replace-russian-air-defence-systems. Accessed on 12 May 2018.

6 https://www.armytechnology.com>indian-army-receives-first-upgraded-schilka. Accessed on 12 May 2018.

7 https://www.defenseworld.net>indian-army-set-for-vshorad-tender-negotiations.Accessed on 13 May 2018

8 Modernisation of Army Air Defence, Indian Defence Review, Vol 29.3 Jul-Sep 2014.

9 https://www. indiandefencereview.com>induction-ceremony-of-akash-weapon-system—into-army-air-defence. Accessed on 12 May 2018.

10 https://livefirstdefence.com>what-india's-new-qrsam-looks-like. Accessed on 13 May 2018.

11 https://www.en.m.wikipedia>qrsam-missile. Accessed on 12 May 2018.

12 https://www. thebusinessline.com>army-to-get-MRSAM-by-2020. Accessed on 12 May 2018.

13 https://m,facebook.com>list-of-radars-under-development/production-by-drdo/bel. Accessed on 13 May 2018.

Conclusion

This book is an humble attempt to present to a discerning reader a commentary on some relevant issues concerning the ground based air defence scene in India. In essence, the book highlights the various nuances of the air defence battle and the multiple challenges faced by the valiant 'vayu rakshaks' in putting up a viable counter to the instant air threat by the potential adversary(ies).

Driven by the eternal cause-effect cycle where the initiative rests with the prosecutor of the air threat, the defender realised way back in the Seventies that protecting each and every VP with terminal weapons was not only futile but also impossible. Soon was born the concept of layered-and-tiered employment of air defence leading to integrated air defence systems (IADS). From this emerged the concept of employment of GBADWS that are relevant today.

This book has brought out how the air defence control and reporting system (ADCRS) is the lifeline of air defence battle and why this all important link must be very strong if GBADWS need to perform optimally. The nuances of Air Space Control (ASC) anchored on the twin challenge of providing maximum freedom of action to the multifarious users of air space with minimum restrictions, as also, to avoid fratricide at all costs has been explained.

The unmanned revolution finds a special place in this book in three different contexts, namely, 'an attacker's delight, a defender's nightmare' and the latest developments in the emerging concept of manned and unmanned teaming or MuM-T.

A strong case is made for the ballistic missile defence capability for India and explains how the 'Programme AD' is all poised to realise this capability provided the various imperatives associated with operationalisation of this capability are addressed.

Looking to the future, the chapter on stealth makes a cautionary call to say that the stealth threat has already arrived at our doorstep and we need to acquire the capability to deal with the same, sooner than later.

The book identifies the grave challenges involved in sustaining a vintage and obsolescent inventory of GBADWS and highlights a possible way out to deal with this challenge. Towards the end, the relevance and special importance of simulators in the GBAD scenario has been highlighted.

Cumulating the major recommendations of all the foregoing chapters, the book presents the fruit of research by identifying the future needs of ground based air defence today.

It reiterates the need for an institutional mechanism for timely and regular update of the air threat as well as an exercise to identify and prioritise the VAs/VPs at the national level from time to time. Besides, the specific action points related to realising the need in each type of weapon and support system has been identified.

Index